RETURN TO ELKINS CREEK

RETURN TO ELKINS CREEK

A FISHING ODYSSEY

EVAN PEACOCK

UNIVERSITY PRESS OF MISSISSIPPI / JACKSON

Books by Evan Peacock

Kudzu on the Ivory Tower: From the Backwoods to an Academic Career in the Deep, Deep South. 2021. Borgo Publishing, Tuscaloosa, Alabama.

Exploring Southeastern Archaeology: Essays in Honor of Samuel O. Brookes (edited with Patricia Galloway). 2015. University Press of Mississippi, Jackson.

Time's River: Archaeological Syntheses from the Lower Mississippi River Valley (edited with Janet Rafferty). 2008. University of Alabama Press, Tuscaloosa.

Mississippi Archaeology Q & A. 2005. University Press of Mississippi, Jackson.

Blackland Prairies of the Gulf Coastal Plain: Nature, Culture, and Sustainability (edited with Timothy J. Schauwecker). 2003. University of Alabama Press, Tuscaloosa.

The University Press of Mississippi is the scholarly publishing agency of the Mississippi Institutions of Higher Learning: Alcorn State University, Delta State University, Jackson State University, Mississippi State University, Mississippi University for Women, Mississippi Valley State University, University of Mississippi, and University of Southern Mississippi.

www.upress.state.ms.us

The University Press of Mississippi is a member of the Association of University Presses.

Publisher: University Press of Mississippi, Jackson, USA
Authorised GPSR Safety Representative: Easy Access System Europe - Mustamäe tee 50, 10621 Tallinn, Estonia, gpsr.requests@easproject.com

Library of Congress Cataloging-in-Publication Data

Names: Peacock, Evan, 1961– author
Title: Return to Elkins Creek : a fishing odyssey / Evan Peacock.
Description: Jackson : University Press of Mississippi, 2026. | Includes bibliographical references.
Identifiers: LCCN 2025047189 (print) | LCCN 2025047190 (ebook) |
ISBN 9781496862778 hardback | ISBN 9781496862785 trade paperback |
ISBN 9781496862792 epub | ISBN 9781496862808 epub |
ISBN 9781496862815 pdf | ISBN 9781496862822 pdf
Subjects: LCSH: Fishing—Anecdotes | Fishing—Mississippi—Anecdotes
Classification: LCC SH441 .P346 2026 (print) | LCC SH441 (ebook)
LC record available at https://lccn.loc.gov/2025047189
LC ebook record available at https://lccn.loc.gov/2025047190

British Library Cataloging-in-Publication Data available

For Josie,
who understands

CONTENTS

RETURN TO ELKINS CREEK

Something Fishy

Those with less cast from the shore,
Aiming for the middle,
Jealous of the ones who go,
In boats beyond their reach,
While in the boats, the few with more,
They motor out a little,
And there in regal splendor throw,
Their lures toward the beach.

I

LEADER LINE

We ask a simple question and that is all we wish . . .
Are all fishermen liars or do only liars fish?
—WILLIAM SHERWOOD FOX

Fishing! It gets in the blood. It gets you out of bed way too early. It's pretty smelly. I love it so. Like alchemy, fishing is a strange mix of seemingly incongruous ingredients. Unlike alchemy, the end product really is gold, if happiness be golden. For fishing brings joy of the purest sort, and has for a long, long time. Evidence for the pursuit may date back as much as 780,000 years.[1] The first known fishhooks, made from shell, date to about 42,000 years ago.[2] Distant ancestors, no doubt. Fishing is far more than just a recreational pursuit. It provides a touchstone in changing times, a salve in hard times, and a wholesome pleasure at all times. It keeps one grounded in the things that really matter: the beauty of nature, the warm glow of companionship, the simple joy of living. Fishing is a place one can go to gain distance from the noise of the world. With distance comes perspective, which always is useful. Somehow, fishing is proof to myself of who I really am. The kind of guy you'd like to share a boat with. Can't ask for much more than that.

I have been obsessed with the sport since I was old enough to shoulder a cane pole and make my way to Elkins Creek, a modest stream running close to our old home place in the rolling hills of Choctaw County, Mississippi. We moved there in 1966, when I was five years old. The house was a ramshackle affair, an old, enclosed dogtrot cabin with one functioning fireplace, yawning cracks between the floorboards, and horrible asphalt siding resembling bricks to a very forgiving eye. It was beautiful, in a Grimm's-Fairy-Tales sort of way, if a tiny place for two adults to raise seven kids. Given that various cousins and other stray-pup acquaintances lived with us at one time or another, it was a small place indeed. But it did have its charms: a melodious tin roof, massive old oak trees, a somewhat askew yet strangely dignified wooden barn softly framed in leafy green pokeweed at the end of a well-trodden dirt path. And the setting was magnificent: miles from anywhere, no neighbors within gunshot, beautiful fields of yellow broom straw, green pastures, deep, leafy forests, and that lovely, lovely creek just a few minutes' walk away. I got to fish in that creek. Gold, for sure; twenty-four carat, shiny and bright.

As I graduated from cane poles to rod-and-reel fishing, I fell in love with rigs and lures, at least up to the point where people become dependent upon gadgetry to locate and target the fish. These days, a fully loaded bass boat has electronics far superior to what initially got us to the moon. I can barely operate a cell phone. But basic gear is inexpensive enough so that almost anybody can join in. Fishing, especially freshwater fishing, is an art, but as with any art, the product is partially dependent upon the tools employed, and the history of development of those tools reflects a fascinating mix of ingenuity, stubbornness, and serendipity. Casting a line has been a practice for a very long time. The rod-and-reel likely was invented by the ancient Chinese about seventeen hundred years ago![3] And the technology has been evolving ever since. With great reverence, I include here some stories about those creative entrepreneurs who bucked convention to create affordable marvels to the betterment of us all.

Fishing is a great leveling enterprise, enjoyed by people from every walk of life around the globe. YouTube[4] abounds with videos of people in less-developed countries using rod-and-reel combos to cheerfully haul enormous catfish (and species of more arcane classification) out of modest pools or flooded fields. Especially when they are biting, political differences and ideological barriers fade in the face of common humanity, shared experience, and friendly(ish) competition. As Iris Murdoch said, "Fishing seems to be the great occasion when all manner of contradictions reconcile themselves." Herbert Hoover put it more succinctly: "All men are equal before fish." We all have become too inured to stridency and discord in the world. We could do a lot worse than to wet a line together now and again. Hoover also remarked, "The gods do not deduct from man's allotted span the hours spent in fishing." Based on that dictum, I may live forever.

It is a good time to be an angler. It is a good time to be alive. As were those seemingly endless summer days of long ago when, as a shirtless, barefoot, forgivably ignorant fledgling of backwoods America, I found gold on the banks of Elkins Creek, cane pole in hand, waiting patiently for that peculiar magic of a bite to ensnare me. It ensnares me still. And as fishing and storytelling are inextricably linked, I hope that you, too, may be ensnared by the following recollections, perhaps to the point where you decide to go out and wet a line yourself. And may you live forever in consequence.

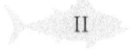

II

HOOKED

Fishing is not an escape from life, but often a deeper immersion into it.
—HARRY MIDDLETON

I learned to fish at a different time, in a different world, when the poorest among us in the Deep South still took our water from holes in the ground and high school was considered an optional step along the career path. There were a lot of mouths to feed in our household: Mom, Pop, and a passel of boys, Dennis, Bennie, Hardy, Glenn, Robert, me, and Andy. Much of our food was wrung from garden plots laboriously broken with an awkward gas-powered tiller and maintained with well-worn hoes and a lot of sweat. Picking was a frenetic season, as lots of things came in all at once. Mom spent long hours processing and canning, the resultant bounty lining up in ranks of glass jars on homemade pine kitchen shelves. Peas, beans, corn, okra, tomatoes—a beautiful display promising a steady supply of nutritious, if mushy, vegetables throughout the long winter months. Meat was obtained from the chicken pen, hog killings, or from the wild, with the occasional exception of boloney thin-sliced from log-like tubes at Henderson's store in French Camp. If we shot something, we ate it, be it deer, squirrels, rabbits, raccoons, beavers, quails, doves; you name it. Once, we even ate a hawk, which cannot be recommended from either a legal or a culinary perspective. There were a few exceptions to this general rule, including snakes and possums. We had our standards, Henderson's boloney notwithstanding.

We cut wood for heat, standing well back as Pop cursed the stubborn chainsaw, darting in to grab logs as opportunity afforded. We loaded the pickup bed with heavy lengths of trunk, slipping and sliding in the winter mud, struggling to keep a grip with cheap, wet, brown cloth gloves when the wood was coated with ice. All of that would have been easier during some other season, but long-term preparation was not Pop's strong suit. The only time he cut more than two or three days' worth of fuel was just before Christmas, so that he could spend about two weeks feeling very merry indeed. Once the truck was unloaded in the front yard, we busted the logs apart with sledgehammers, wedges, single-bitted axes, and pure muscle. When an axe head grew loose, sixteen-penny nails were driven into the top of the handle to tighten it up. The protruding shanks were hammered over to form hooks across the sides of the eye, a safety feature designed to keep the axe head from flying off and diminishing the sibling count. One species of oak smelled just like cat mess when split, adding to the joy of the experience. For kindling we used scraps obtained from the box factory outside of French Camp, long, thin strips of dry hardwood that flew like dangerous shrapnel when chopped in half with a hearty swing of the axe. When those sticks ran low, Pop put corn cobs in a paint tray on the back porch and soaked them in gasoline. It made for an interesting smell when he started the fire on winter mornings. Pop was a mad genius in that kind of way.

Some extra cash was generated by cutting small pine trees for pulpwood. They had to be small because we had no winch. We handloaded the trunks onto the protesting pickup to be hauled, bed nearly dragging, to the mill outside Weir. The average load returned about fifteen dollars. Those were the days when "a dollar's worth" was the standard unit at the gas pump, credit lines were stretched well beyond thin, and a nickel found in the crevice of a car seat was an appreciable treasure. In short, life as we knew it required work, and lots of it, often of the most exhausting kind. But in the welcome interludes between chores, the creek beckoned and therein lay our reward. It was a beautiful, sinuous ribbon running nearly clear where ice cold springs mixed the waters, turning cream-coffee brown during the larger spring floods. It meandered through picturesque hardwoods, with bluffs just high and steep enough to encourage the imagination. It was rich with life.

Especially during the spawning season, if the moon was full and the water was low, you could *smell* fish, even from our house. The smell of that creek coming in through the screen of an open window on warm currents of nighttime summer air was memory in the making. How I would love to smell it again.

We grew up wild, free, and only nominally washed, adding to our riches. I am sure that Mom and Pop worried about snakes, especially as we kids went barefoot for at least four months out of the year. But snakes did not signify; neither did poisonous plants, blood-sucking parasites, treacherous deadfalls, or any other of the innumerable hazards of the deep woods. That creek was simply irresistible to a slew of rugged country boys, the older ones of whom remembered far too many rent-related changes of address prior to inhabiting our country abode. We saw snakes, of course: fat, stinking moccasins, beautiful but aggressive copperheads, chicken snakes, corn snakes, king snakes, hog-nosed snakes, and many others. On one occasion each, a rattlesnake and a coral snake were spotted, the latter swimming upstream and identified by its signature color bands. One stepped high and made a lot of noise while in the woods. Numerous other interesting lifeforms inhabited or were drawn by the water, from annoyed crawdads to torpid turtles to wary beavers. The turtles would line up on the trunks of trees that had fallen into the creek, warming themselves where the channel was wide enough for sunlight to fall. When signs of danger reluctantly entered their reptilian brains, a one-step command sequence—LET GO!—was initiated, and they would fall in sequence into the drink like train cars derailing from a bridge. *Plorpplorpplorpplorp!* Beavers saw you before you saw them, giving notice by the rifle-shot slaps of their tails on the water. Their slides abounded, glisteny lick marks punctuating the leaf-strewn banks. Deer crossings were marked by muddy trails and raccoons left little paw-glyphs on many a tiny sand bar. Our own footprints similarly marked those bars until the next rain raised the creek, temporarily erasing the record of passing of all creatures. Such were the transient signatures of life in our insular little world.

We spent endless hours ranging the well-worn creekside trails. Massive grapevines provided swings we fearlessly used to cross from one bank to the other. Fallen trees made the most romantic of bridges for cowboys, ape-men, and other characters inspired by the pulp fiction section of our eclectic little cobbled-together library. Small islands guarded the outer bends of meander loops, worthy headquarters for skinny pirate scalawags in tattered red do-rags. Mysterious, deep pools beckoned beneath the exposed root mats of somber beech trees. Shallow stretches of creek bottom sported old cork-aperture bottles, small pieces of petrified wood the color and thickness of Captain's Wafers, and other treasures readily collected by fingers chilled to the bone in the spring-fed water. And best of all, better than almost anything ever, was the fishing.

Our fishing apparatus was simple and partly self-manufactured. We cut poles from a little canebrake deep in the woods east of the house, chopping the stalks off near the ground and trimming the leaves with machetes Pop fashioned from old lawnmower blades. The handles of those machetes were made out of aluminum beer cans, a type of raw material for which there always was a ready supply conveniently located around his shop. Pop melted the cans down in a forge and cast the handles in a homemade mold, attaching them to the heavy blades with carriage bolts run through the sides. Those machetes hurt one's hands like hell, but we did not complain. Some things are just meant to be borne.

Back at the house, the poles were cut to length and suspended for several weeks from tree limbs, with bricks or pieces of cinderblock tied to the bases so that they dried straight and true. Finished and outfitted, they leaned like so many teepee struts against the outer-most oak tree of the front yard. Line came on spools of whatever was cheapest at Henderson's. There we also bought bright red-and-white bobbers with recessed catch wires that rose like mantis claws when the central, hard plastic cylinder was pressed upward. We did not call them "bobbers," of course; we called them "corks," in the same generic way that all soft drinks are called "Cokes" in the South. We did sometimes use actual corks, found in old bottles cached in barns or sheds on the farmsteads of our many backwoods acquaintances. One makes a lot of friends riding the school bus for an hour each way. Mr. Braswell, an elderly gentleman with an astonishing hunchback, lived about half a mile down the gravel road. He routinely gifted us with hand-carved wooden floats, the manufacture of which was his dearest hobby. We always accepted them graciously, then discarded them as soon as the coast was clear, as Mr. Braswell's hardwood floats became waterlogged and consequently nonfunctional in short order. But the red-and-white plastic bobbers were iconic, and we were sure that something about the color configuration served to hypnotize fish, convincing them that dinner was served.

Somewhere, Andy and I picked up a little brown cardboard box of Pfleuger hooks, the name of which conjured a certain magic that we were sure led to even greater success. We always pronounced the silent "p" as part of the incantation: Puh-*flueger*! Sinkers (or "weights," to use another generic term) were easily obtained. The tin roof on our house had been affixed using lead-capped nails. The caps spread upon hammering, providing some protection against leaks. As the nails aged, and with seasonal expansion and contraction of the sheets of tin, the caps came loose and gradually slid down the roof to collect in the drip line around the house. They looked like little gray berets. We collected them by the dozen and fixed them to our lines by folding them double and clamping down with our teeth. We seldom had storebought stringers, so we improvised by cutting forked sticks out of bankside saplings and running the stems through the gills of our catch. Clumsy, but functional, up to a point. On occasion the whole family would embark on a major fishing expedition. One such event did involve a storebought stringer, a long, yellow one with a sharp metal point on one end and a stout metal ring on the other. We caught over a hundred fish that day. The filled stringer was longer than I was tall. I remember it as being as long as Andy and I put together, but one has to correct for decadal memory drift. Regardless, it was a real mess of fish. That was a good day, especially as it ended with a marathon fish fry in the back yard.

Gathering bait was almost as much fun as fishing. Earthworms were our go-to. Following heavy spring rains, twiggy lines of wrack ran like agricultural terraces under the oak trees in the back yard. A quick scrape across the detritus with a kitchen fork would expose several squirming, brownish-pink tubes. One had to be quick with the business end of the fork to lever the worms out whole before they wriggled down into the ground. Worms and some enclosing, protective dirt went into an empty mayonnaise jar for transport. The worms left an oddly comforting smell on one's hands that lingered for hours, practical hygiene not being the strictest of requirements in our household. Robust arguments ensued concerning whether the worms should be threaded or bunched on the hook. I was in the threading camp, as it took the turd tappers—tiny bream too small to be caught— a longer time to worry the bait off. A turd tapper was signified by the cork doing a characteristic sideways swirl, never quite going under. There were a lot of well-fed turd tappers in Elkins Creek in those days. Worms left on hooks between expeditions hardened into dry, lacquer-like cases that had to be broken off like mummy fingers before fresh bait could be emplaced. Experiments with rehydration resoundingly failed; the Elkins Creek clientele were too sophisticated for such sec-ondhand grub. Not so the dogs; hook points had to be left stuck into corks lest the greedy hounds impale themselves trying to consume any remnant bait. We learned early not to let Robert carry the worm jar, as he was an awkward creature in those days. One expedition came to an abrupt halt when for some reason he was delayed in joining us. Racing to catch up, he failed to notice the jar, accidentally kicking it into the water. The fish must have worshipped him as a god. We were somewhat less reverent on that particular day.

In high summer, when the ground was too dry for digging worms, we used a hunting knife to poke air holes in the metal lid of a can-ning jar and wandered the field by the house gathering grasshop-pers. I had no fondness for those creatures, with their horny legs and nasty habit of expectorating tobacco juice onto one's hands. Crickets worked better as bait, but they were found only in limited numbers beneath old boards or other waste around the house. More beauti-ful, but completely useless for fishing, were the "graveyard crickets," as we called lubber grasshoppers. Startlingly black, with bright red or yellow abdominal stripes, they traveled like ominous biker gangs from one patch of woody vegetation to another. We tried them many times, with never a bite to show for our efforts. Like bikers, their colors clearly conveyed a stern message: mess with us and die. The fish took note.

One year the older brothers dug a large hole in the field between our house and the creek to serve as a storage facility for dozens of box turtles collected in the woods. The ensuing reptile races being not nearly as exciting as one might imagine, we eventually lost interest in the enterprise, whereupon Mom decreed that the contestants be returned to the wild. The hole soon filled to the brim with rainwater and became home to astonishing numbers of minute crawdads, progeny of their larger counterparts whose goofy looking mud chimneys dotted the field each spring. Their little babies were readily scooped out of the milky brown water with a kitchen strainer and stored in a jar, where they darted to-and-fro in an accompanying sample of their home environment. The bream loved those little guys, and we made the most of the situation before a man hired to break up the field hit the hole with his tractor tire and very nearly turned over, at which point Pop made us fill it in. It was good while it lasted.

Hardy, the most adventurous of us all, once decided that what we really needed to use as bait was red wasp larvae. I do not know where he got the idea, but then, it was hard to say where Hardy got any of his ideas, although one suspected that aliens were beaming them directly into his brain. He and Cousin Dwight, who was living with us at the time, undertook the delicate task of knocking the nests down without getting stung. They failed miserably, but did eventually succeed in extracting scores of disgusting little yellow larvae for their pains. The effort was not worth the sacrifice, alas; the Elkins Creek bream, spoiled consumers that they were, turned their refined noses up at wasp larvae. We younger brothers nonetheless were impressed at our elders' derring-do, although we never chose to replicate the venture.

One blistering summer, when times were particularly hard, all of our resourcefulness was called upon in a scene worthy of a Rockwell painting. All the good cane had been harvested from the brake to be used as bean poles in the garden. We were out of line and store-bought bobbers, with no money to buy more. Worms were hiding deep beneath the sunbaked ground surface, crickets were vanishingly scarce, and even the grasshoppers seemed to be on strike. Undeterred, Andy and I cut saplings for poles, knotted together bits of old badminton racket webbing for line, tied on pieces of broken sticks for floats, attached the legendary Pfleuger hooks, clamped little lead berets down with our teeth, and *voila*! We had our fishing gear! Using wild chives, we yanked tiger beetle larvae (a.k.a., "chicken chokers") out of their cylindrical lairs in the ground and impaled the vicious little bastards on the hooks, avoiding their demonic pincers while trying not to stab ourselves in the process. I do not remember catching anything with those rudimentary outfits, but the episode speaks volumes about our dedication to the cause.

Favored fishing holes were granted names signifying some geographic or cultural feature or honoring the discoverer. The Bridge was one such locale. Other spots had more bites, but on the Bridge, you could sit with pole in hand, legs dangling, enjoying the shade of the over-branching hardwoods while monitoring the cork. If you held your pole still enough, blue-green dragonflies would alight and help with the monitoring. Or you could try to assassinate minnows with a B-B gun if nothing larger was biting. The Bridge itself was a rattly wooden structure that produced a kettle drum roll whenever a vehicle passed over. At first, a well-illustrated edition of *The Three Billy Goats Gruff* kept me from venturing underneath, but it was not long before I began taking up a familiar station on a soft sand bank there, plopping my line down beside the sturdy piers where leviathans surely lay in wait. If a vehicle passed over, you were inside the kettle drum. Brr-rrr-rrr-*room*! Fortunately, that was a rare occurrence. Creosote, sunbaked gravel, bottomland forest, and creek water contributed to a heady scent that I still can recall. There is no reckoning the number of hours we spent on the Bridge. They were some of the best hours of my life.

The Raft was a piece of excellent, albeit inadvertent, fish cover created when the older brothers' homemade conveyance went to the bottom after a regrettably brief maiden voyage. For many of their experiments, they employed younger brothers as test subjects. Elkins Creek flows into Poplar Creek, which flows into the Big Black River, which flows into the Mississippi River, which empties into the Gulf of Mexico. I suppose they wanted the glory of that expedition for themselves. Glory there was not, but no one drowned and the fish were happy with their new shelter. That is what one calls a happy ending.

The Log was a semi-submerged tree trunk that diverted water in a bend, creating a deep hole that never failed to produce fish if one were patient. A particularly deep, small hole developed beneath the exposed roots of a beech tree just upstream from Bennie's Island. One could stand on the root mat and try one's luck in any of the myriad little openings spread like woody filigree upon the water, each a potential *McElligot's Pool*. How we would have gotten anything big up through that root mat was a matter to which we gave no consideration. Mother's Hole was a wide, placid expanse where a mass of floating debris and yellowish-white creek foam had piled up behind a log. While that feature lasted, the fishing was unbelievably good; the cork would not even pause before it went under, the bait greedily consumed by some burly panfish.

Far to the west, almost to Highway 413, was the Beaver Dam, a fine, still pool bordered on the downstream side by a scenic stick and mud barrier erected by those busy little critters. By far the most consistently excellent fishing spot on the creek, the Beaver Dam was quite a trek for little legs, so we used to take a shortcut across the pasture fronting our property, inevitably catching the attention of the sour-tempered Brahma bull whose territory we were invading. It is hard enough to clear a barbed wire fence when you are in the neighborhood of five feet tall; doing so with a charging bull closing in behind you, with a cane pole in one hand and an open jar of bait in the other, is quite the challenge indeed. But we were persistent little cusses. And the fish were waiting.

The creek was home to an impressive array of fish species, most of which we considered edible if they could be extracted via hook and line (or M-80s, a one-time endeavor that doubtless was Hardy's idea). Bream ("brim") was a general term for all kinds of beautiful sunfish: green, bluegill, warmouth, redear, spotted, longear, pumpkinseed. Warmouth was the only kind we distinguished, calling them "goggle eye," a folk term that we eventually changed to "googly eye" because that was hilarious. On the short nonedible list were chubs, which we called suckers, and shiners, both of which were rare catches in any case. Channel cats and mudcats were common, if never of particularly impressive size. That did not matter much, as we observed no size or catch limits; indeed, the very idea of such restrictions was foreign to us. If it was "big enough to make a grease spot in the pan," it was big enough to clean and eat. Despite our best efforts, we never seemed to dent the population. A new crop of turd-tappers always was ready to bloom.

Sometimes the cork would behave strangely, moving erratically around without submerging. We knew what that meant; a large crawdad was helping himself to the bait. It always was dramatic to lift those weighty creatures up out of the water, where they would give a look both alien and disdainful before releasing their impressive claws and dropping with a loud *plunk* back into the watery depths. We never tried to catch and eat those crawdads, despite their size; nor did we keep the smaller ones we caught by "fishing" the chimneys in the field. I do not know why we did not eat them, given that we unabashedly ate pretty much everything else that walked, swam, flew, or crawled (except possums and snakes; standards, you know). But in retrospect I am grateful, as we kids did consume a few small crawdads grabbed out of tiny streams in the modest hills north of the house. Imagining ourselves hardy pioneers, we would pull the tails off and "cook" them with kitchen matches, invariably burning our fingers in the process. They tasted terrible, and I always felt sort of odd afterwards. Allergy, it turned out, which extends to shrimp, lobster, and probably any other claw-bearing crustaceans. People commiserate, which I appreciate. But pass the tater salad, please. And if you are allergic to tater salad, may the Good Lord have mercy on your soul.

A haul of fish is called a "mess" for a reason. Expert fish cleaners, we were not. Filleting was an unknown skill, and we might not have employed it in any case, as it is inherently somewhat wasteful of meat. Bream were scraped with a spoon to remove the scales, an unpleasant process that sent small, translucent discs flying in all directions. Many were found later clinging to one's skin or on some supposedly pristine kitchen surface, where they had dried and assumed the shape of little taco shells. Scaling was followed by beheading and gutting. Heads and guts went to the hogs, who accepted them with greedy delight, devouring the boney remains with a horrific crunching sound while wiggling their curly little tails in compliment to the chef. Catfish were nailed to an oak tree so that the skin could be stripped off with pliers. We conked them on the head with the hammer first, but it still was a gruesome operation.

Most fishing expeditions were relatively short, as chores always called, so that we would bring back just a few fish at any given time. Lacking a big freezer, and plastic storage bags being outside of the family budget, we would put three or four cleaned fish in a bowl of water and stick that in the small freezer atop the refrigerator, tail fins sticking out over the edge of the bowl like so many Chinese fans. It did not take long before things became crowded up there, at which point we would go on another, more prolonged expedition while the prior catch thawed and then have fish for supper. As was always the case, Mom caught the brunt of the work, if catching and cleaning are not counted (and catching never counted). She would melt big, serving-spoon dollops of shortening in a large cast-iron skillet, roll the fish in seasoned cornmeal, and fry them up along with enormous quantities of golden-brown hushpuppies. The result was delicious, salty goodness, satisfying in the extreme. The tail fins were crunchy, like potato chips but more gratifying in a primeval kind of way. Bones were always a danger, and more than once the feasting was interrupted as Mom helped one of us kids dislodge a rib from his throat. The whole undertaking was messy, fun, and highly rewarding, especially when beguiling summer Twilight spoke to us greasy young adventurers as we embraced her with love, lightning bugs winking approval in the heavy, humid air. Thank you, Mom. Thank you, Twilight. Thank you, Elkins Creek!

Fishing was a passion that burned brightly in my heart. I monitored the cane brake for pole growth. I gazed out the window when the school bus crossed old wooden bridges, imagining what it would be like to fish the mysterious waters below. In a field beside the elementary school was a modest, square, grass-lined pond that in my mind was full of hungry catfish and angry bream. Turned out to be an open air sewage pond. Well, crap. Hankering for fishing got me into trouble early on. One fine summer day when I was about eight, I decided to hit the creek to see what was biting. I dug up a few worms, grabbed a handy pole, and set a course for the Log. They were not exactly tearing it up, but eventually I caught two catfish, just like Sam Gribley in *My Side of the Mountain*. I proudly headed home, only to find that I was the one in the frying pan. In my eagerness to get my line in the water, I had neglected to tell anyone where I was going. My punishment was dire in the extreme: perhaps motivated by my catch, my brothers set out for the creek, poles in hand. I was forbidden to join them! Devastated, I stood on the porch beside Mom, watching as they went. Her better judgment yielding to her soft heart, she relented upon seeing the stricken look on my face. "Go on, then!" she said in an exasperated tone of voice. I flew off the porch and raced to join the expedition. I cannot remember if I paused to say thanks; I hope

I did. She truly was a mom for the ages. Naming a fishing hole after her was an insufficient, if high, honor.

In the many years since those cane pole days, I have fished a lot of different bodies of water. I have caught thousands of fish. Why, for a while I even had a sparkledy boat! But nothing has ever matched the simple happiness I found on the banks of Elkins Creek, a middling stream in the rolling hill country of Choctaw County, Mississippi where the bream were belligerent, the woods cool and inviting, and shirtless, shoeless brothers welcome companions on summer days as yellow and rich as storebought butter.

III

FIRST CAST

I go fishing not to find myself, but to lose myself.
—JOSEPH MONNIGER

The year I turned fourteen, two events of particular moment took place in my life: I used a telephone for the first time, and I learned how to use a rod and reel. The telephone experience was weird. The rod and reel experience, sublime. My love of fishing was about to be catapulted to a whole new level.

I experienced this breakthrough thanks to my good buddy Mike. Mike was an interesting guy, as dour and stubbornly independent as an unneutered tomcat, yet fast and true as a friend. The only boy among five children, he stepped early into an adult's shoes when their dad died far too young. The siblings were girls, two of whom were raw, vivacious, approximately my age, and something of a fixation in my young adolescent fancy. In the strangely cryptic yet clear body language that only teenage girls possess, they made it known that I was near the bottom rung on the ladder of potential boyfriends—surely, I was at least one step above Danny H.?—but we all had a great time together. Fancy unrequited is fancy, nonetheless.

Mike and the recently nubile sirens lived with their mom in a ranch house way out in the country outside Winona, where the North Central Hills and the Loess Hills merge, a rumply-bumply earthen fabric slowly being ironed flat by gravity and the elements. With considerable help from the human element, one might add.

A century-plus of poor farming practices produced a sad legacy of erosion, the dissected ridge systems looking like they had been raked by giant dinosaur claws. Driving the serpentine gravel roads was a treacherous enterprise, as one moment of carelessness would mush a tire into the sandy shoulder and drop-suck the vehicle right over into a hungry roadside ravine. I once was riding shotgun in a van that got swallowed up by such a ravine. It was not an experience I care to repeat. The entire landscape seemed ready to fly apart, held together only by the omnipresent kudzu. Suffocating trees rose out of the green mass like saguaro cacti, or leafy sock puppets gesturing mutely to one another in a play whose ending was not guaranteed to be a happy one. It was not wilderness, quite; too many lost hunting dogs and discarded beer cans in the ditches to make such a claim. But if it was not wilderness, it was just this side of it, as ol' buddy Mike was just this side of domesticated.

Becoming man of the family at such a tender age perhaps exacerbated an innate headstrongness in Mike that manifested in unfortunate ways, such as the time he sought to encourage a reluctant fire using gasoline housed in a milk jug. Upon contact twixt fluid and flame, fire instantly lunged like a starving carnivore out of the fireplace, up the amber stream, and into the jug that Mike wisely discarded as soon as was humanly possible. Not so wisely, he discarded it into the living room, although his options at that point admittedly were limited. To my knowledge, that is the only house he ever burned down, but it would not surprise me to be corrected on that score. Mike was wiry, knobbly of elbow and knee, prideful, taciturn, and tough as homemade jerky. He seldom laughed, but when he did, everyone in the room laughed with him. In rare moments of complete abandonment, he would break into the "Chicken Strut," a dance move of cosmic absurdity. Just as no one could have played Conan the Barbarian as did Arnold, never has anyone done the Chicken Strut like Mike. It was his one signature move in an unforgiving world, and he made the most of it. I loved that guy; I hope he struts his chicken still.

One fine summer day Mike and I went to a little farm pond, thus beginning my undying love affair with little farm ponds. They are fascinating places, rich oases of life harboring everything from fish to snakes to muskrats to mussels. This one was perhaps half an acre in size, water the color of caramel, surrounded by brown swaths of dry pasture sutured together with zigzagging red cattle trails. There was no shade, but that did not matter as both of us were coppered by habitual exposure to the sun. "Brown as an Indian" was what they called it in those days. I knew nothing wrong about it then; rather an honor, in fact, given that we all proudly claimed Native American ancestry, a popular myth only grudgingly giving way to the realities of affordable genetic testing in these astonishing times. Indians, we were not. Proud of our outdoors skills, we were. Justifiably so. I was about to add to mine.

I have no idea to whom that pond belonged, but people were kinder about such things in those days, customs embodying logic born of necessity in a state where the Depression had hit hard. Under that system, hunting or fishing on someone else's land almost always was accommodated, often in the most warm and welcoming way, as long as you asked first. Such generosity traditionally was rewarded with a hindquarter or a mess of fish. If it was a *really* good fishing hole, the fish would be delivered cleaned. And, of course, people were welcome on your land in turn, if you had any. And if they asked first. Things changed when rich folk in Jackson and Memphis began buying up enormous land holdings that became the modern-day equivalent of the King's Forest, and that *still* did not stop them from trespassing, cutting fences, stealing dogs, firing high-powered rifles from blacktop roads, and otherwise acting the fool. Which, of course, brought out the meanness in us. With which we only victimized one another, conditioned by that peculiar Southern mindset that always seems to give rich folks an out, except that we trespassed freely on the King's Forest. Somehow, that fit into our logic system as well, which was damned convenient.

Mike had taken it upon himself to teach me the fine art of rod-and-reel fishing. A greater kindness I never have been done. We both were sporting Zebcos, courtesy of Mike. Almost all freshwater fishermen cut their teeth on Zebcos, sturdy little closed-face spinning reels that represent a typically weird American success story.[1] A perennial problem with any sort of fishing reel is backlash. Backlash is to anglers what cell phone junkies are to theatergoers: exasperating distractions best excised from the Big Book of Creation, or whatever it is

God keeps His ledgers in these days. What happens is that, following release of the line, the reel keeps spinning for longer than desired, so that excess line instantly piles up into a "bird nest," bringing all fishing activity to a halt while the mess laboriously is sorted out. Inevitably, one's fishing partner will loudly exult in the greatest angling experience of his or her life, lunkers feeding with piranha-like voracity on anything that hits the water, whilst Operation Blankety-Blank Bird Nest Elimination is underway.

An inventive Texan watchmaker named Jasper R. D. Hull studied the situation and came up with a more-or-less backlash-free reel inspired by "watching a meat counter clerk pull string from a fixed spool."[2] Finding a marketer turned out to be surprisingly difficult, suggesting that Mr. Hull was perhaps a bit too ready to share his meat counter clerk story. Options growing thin, he finally approached the Zero Hour Bomb Company, whose sole product at that time was electric time bombs used in fracturing rock for oil extraction. Hull's proudly exhibited prototype was "a piece of plywood with nails arranged in a circle and wrapped in line. Attached was a coffee-can lid that could spin."[3] Apparently, that was just the needed enticement for a bunch of rock-fracturing time-bomb executives. It was a match made in Heaven. Their first fishing reel, the "Standard," hit the market in 1949; sales boomed far more than had the bombs, and a legend was born. In 1956, a reel was sent to President Eisenhower at the request of a friend and fellow fishing nut. Upon seeing a package labeled "Zero Hour Bomb Co," a different kind of nut was suspected by White House security personnel, who plunged the parceled gift into a tub of water and called the bomb squad.[4] That debacle led to the company's catchy name change later that year. While anglers who can afford to do so usually graduate to fancier, more expensive gear, Zebcos remain a reliable workhorse of the industry.

The lures Mike and I were using were plastic worms of a kind that to this day you will find stapled to white cardboard backings hanging in dusty ranks in the less traveled aisles of country stores. Ghastly, flesh-pink worms six inches long that sport not one but two hooks, shiny golden ones of slender gauge, one anterior and one posterior, shielded behind thin wire projections and linked along the oddly flat belly of the worm by heavy fishing line knotted to the hook eyes.

Atop the frankly repulsive worm is a series of colorful little red plastic beads, in the middle of which is a flashy gold-colored blade. The whole ensemble looks like something that might have crawled from underneath the wreckage of an aluminum foil–shielded spacecraft in a 1950s B movie. Having only ever used live bait up to that time, I looked upon the unlikely artifact with a mixture of analytical interest and haughty disdain. What sort of idiot fish would attempt to eat such an abomination?

How little I knew. I later learned that the lure I initially scrutinized with such mockery was, in fact, the legendary Creme worm.[5] Nick Creme was a machinist in Akron, Ohio in the 1940s. Not surprisingly, given his dull vocation and dreary industrial setting, Mr. Creme direly needed an outside recreational pursuit. Fishing fit the bill. At that time, artificial worms were made of rubber, and anglers generally were less than impressed with their performance. Creme thought that plastic might work better. The DuPont company was in nearby Cleveland, and DuPont had invented nylon, which revolutionized the stocking industry, so they already had a leg up, so to speak. A lab tech at DuPont, whom one suspects was bartering for access to good fishing holes, gave Creme some chemicals to take home. There, he and his wife Cosma—Cosma Creme!—"cooked up mixtures of polymer, pigments and oils. They lugged the glop to the basement and poured it into a mold he'd made from a steel model of a real nightcrawler. After months of trial and error, they produced a plastic worm that looked alive."[6] Convincing anglers to buy his "Wiggle Worms" took Creme some time, although an early clientele were pranksters who liked to hide them in salads at dinner parties.[7] As the worms began to be key to winning tournaments, the real salad days began. Sales took off, leading to a profusion of shapes and frankly hallucinatory colors (cranberry, root beer, ice creme daiquiri, and more)[8] and starting a revolution in bass fishing via the mass production of soft plastic lures.

Mike showed me how to cast by holding down the black plastic button on the butt end of the reel, flicking the rod like a buggy whip and releasing the button at just the right time so that the line was freed and the ludicrous, storebought worm went sailing out in a graceful arc over the water. Click, swish, release, whizz, *plunk!* The lure was retrieved by turning a little handle on the side of the reel. On about my third try, I actually managed to get the worm an appreciable distance out into the pond. I started slowly reeling in and almost immediately felt a small bump through the clumsily conjoined conduit of lure, line, rod, and central nervous system. Curious, I thought. Having reeled in, I repeated the operation: click, swish, release, whizz, *plunk*! Reel; bump. Bumpbump. Bump. After about half a dozen repeats, I began to suspect that something was up. I called out to Mike on the far side of the pond, where his own worm repeatedly was coming up unmolested.

"I think I'm getting bites," I said.

"Are you setting the hook?" he asked in annoyed voice. I immediately intuited what he meant, and on the very next bump I yanked back, hard. *Whammo!* I hooked my very first largemouth bass! With a luck that seldom has abandoned me since, I began reeling in fish after fish, none very big, but that did not matter as I laughed aloud with the pure delight of it all. Where had rod and reels been all my life? Where had *hook sets* been all my life? Mike's lure remained sadly unmarred by bumply little bass teeth as I hauled 'em in. I thus experienced another first: outscoring your fishing partner, a truly integral part of the enterprise. We anglers are a competitive lot, a trait that I suspect conveyed a significant selective advantage upon our hunter-gatherer forebears.

Although I did not realize it at the time, what also began that day was an important education in the complexities of freshwater fishing. Mike had been fishing with rod and reel for years. He and I were using the exact same kind of lure, in the same small body of water. And yet, I caught far more than he that day. Pure luck, of course, but over the years that luck has morphed into a body of experience coupling an eye for water depth, clarity, topographic structure, cover, season, time of day, and weather conditions with considerations of rod length, diameter, and composition, reel gear ratios, line type and weight, lure type, color, and weight, hook type and size, retrieval speed, and other variables. To become familiar with those variables is to evolve into a pretty good angler; to develop what is called a "feel" for the water. One can read, and one can inquire, but the only real way to obtain the feel is to go out and get it. To a soul, professional bass fishermen always have the same advice for those seeking to improve their craft: go spend time on the water. As of that fine summer day with Mike, I never wanted to leave it.

My mission was clear. I set out to obtain my own gear. Money was needed. I hauled hay, shocking my employers by demanding the princely sum of a nickel a bale. I read water meters for Mr. Wesley, jumping in and out of the bed of his pickup truck as he pulled over at meter after meter, flipping the rusty iron lids, wiping scum off the dials, and yelling numbers back while trying to avoid being touched by the inquisitive squiggly things whose homes I had invaded. A dollar an hour for that one. I picked blackberries for a dollar a gallon, obtaining in the process innumerable chigger bites on my tackle box. The pain was all worth it when, at last, I was able to purchase my very own Zebco. I bought some rather heavy line, tied my first clumsy knots through the eyes of whatever cheap tackle I could afford, and began flailing every water body to which I could gain access like a mad schoolmaster proselytizing for corporal punishment. I lost a lot of fish in those days, but negative reinforcement was part of the adventure. How else does one learn that The One That Got Away always out-masses those on the stringer by at least a factor of five?

For the eleven or so non-anglers who may be reading this book, a brief word here on some basic lingo, which I also began to acquire at that time. A consistently excellent fishing spot is a "honey hole." Big bass were in those days called "lunkers." These days they may be referred to as "toads," "pigs," "hogs," or "donkeys." Little ones are called "dinks." Really itty-bitty ones, I started calling "Englebert Überdinks." Sadly, that may have been my finest moment. When a fish hits your lure, it is called a "strike." "Drag" refers to the amount of force needed to pull line from the reel even when the bail, a locking device, is in the closed position. Drag is set by adjusting a small wheel or knob on the reel, an operation that can be executed while a fish is hooked depending on how much weight and pull one feels and, relatedly, the condition of one's heart. If the drag is set too tight, your line breaks; too loose, and the fish will throw the hook. A lunker will "pull drag," taking line off a locked reel with an audible *bzzzzz*, a sound at once terrifying and exhilarating. Catching more than your fishing partner is known as "waxing" the competition. Lifting a fish into the boat without using a net or your hand is called "flipping." Getting "skunked" carries the usual meaning: not catching a single fish, a fate well-nigh worse than death, especially if your fishing partner is spared a similar humiliation. "Doubling up" is when you and your fishing partner each have a fish hooked at the same time. It only counts if both fish are landed, though. "PB" stands for "personal best," signifying the largest caught specimen of any particular species. In the South, weight in pounds is the standard unit employed. Somewhere around the Mason-Dixon Line, length in inches becomes the preferred metric. The Freudian aspects of this distinction are best left unexamined, although a diligent historian might uncover a heretofore unexplored factor leading to the Civil War.

There are far too many lure types to describe in detail, but if you want to fake your way through the topic at the next family cookout, know that there are several general categories of lure: topwater baits, spinnerbaits, swimbaits, crankbaits, jerkbaits, chatterbaits, jigs, and creature baits. These are not mutually exclusive categories, and some lures may be used in combination, but close enough. Topwater baits will be described later. Spinnerbaits have one or more blades that rotate during retrieval, attracting strikes via flash and vibration. Blades come in different forms, the main ones being willow leaf, Colorado, and Indiana, each suitable for different water conditions. Swimbaits are soft-bodied lures that look more or less like little fish of various kinds. They may have paddle tails, ribbon tails, curly tails, fork tails, boot tails, or flat-fin tails, all of which produce a different kind of action during retrieval. Crankbaits are designed to resemble prey like shad or crawdads and have beautifully sculpted and painted bodies bearing one or more treble (three-pointed) hooks. Most have a "bill" or "lip," a plastic extension at the front that causes the lure to descend to depth, partly determined by the speed of retrieval as you "crank" them back; the lip also makes them wobble enticingly. A subset, "lipless" crankbaits, works particularly well in the spring, especially those that have a little internal chamber holding metal beads that rattle during retrieval. Jerkbaits, or stickbaits as they may be called, are slenderer prey-shaped lures, usually billed, that are jerked back to produce an effect like a wounded bait fish. They may be hard- or soft-bodied. Chatterbaits have metal tabs atop the jig head that produce an intense vibration, making them good lures for murky water.

"Jigging" is a general term for repeatedly raising and dropping the lure. Jigs have rubber skirts and usually some sort of plastic body threaded onto a hook weighted with a molded head. Jig heads come in a variety of forms: football, swinging football, swimbait, half-moon, aspirin, banana, egg, glider, gravedigger, dart, bullet nose, and so on. I need a job naming jig heads. Creature baits are lures designed to look like . . . well, like creatures of various types: fish, crawdads, lizards, salamanders, turtles, snakes, leeches, flukes, dragonflies, cicadas, hellgrammites, nymphs, stoneflies, mayflies, wasps, bees. Some look like spiders. Some look like bats. Some look like squirrels. Some look like baby ducks. *Baby ducks.* Some do not look like anything that has inhabited the Earth since the Cambrian period. They are weirdly beautiful. One could make an interesting holiday display with the things. An obvious extrapolation is that bass will eat *anything.* Hopefully this primer has been helpful. Next time Uncle Elwood is holding forth on the topic while demolishing his fourth hamburger, tell him you have been given to understand that black and blue is a good color combination for jig skirts in cloudy water in the spring. That ought to plug his hole.

Eventually, I considered myself expert enough in rod-and-reel fishing to instruct Andy in the operation. We walked north up the gravel road fronting our little piece of land, past green pastures and yellow fields of broom straw, then further north along Old Bankston Road, an old, abandoned dirt trackway that plunges like a subway tunnel into an area of dense, second-growth hardwood forest brightened here and there by aromatic tufts of cedar. Bankston once was a thriving community with stores, churches, and the first successful mechanically powered textile mill in the state, producer of cloth and shoes for the Confederacy before it was burned down by Grierson's little force in 1864. Now it is a ghost town with only the cemetery remaining, out in the deep woods a few miles northeast of our home. And thus do the tides of history rise and fall.

We did not call it the Old Bankston Road; to us, it was the Curtis Road, reflective of some piece of local history about which I remain sadly ignorant. The road itself was an adventure, frequently a very muddy one. The ditches on either side were paved with iron-rich concretions and pale pieces of fragmented branch wood. Some of the pieces were mushy, some semi-solid mineral-clay, some stone; you could *see* fossilization taking place! One never knew what sort of wildlife might be spotted along the leafy corridor. Deer and box turtles were common sightings, red foxes and bobcats rare treats. Sometimes we pedaled in on our sturdy red bikes, it, but even in summertime your tires were likely to become covered in a heavy clay sheath so that you wound up pushing the bike while carrying the gear. Walking was easier, although the horseflies always knew when there was something on the stringer on your way out. Once on the scene, they inevitably decided that primates were the choicer cuisine.

Getting to the pond was another adventure, as it was a long way down the road, then west through the woods until one came to the overgrown little pool, not a quarter of an acre in size. Navigating thick woods with fishing gear was a tricky business, but worth it when that beautiful little pond appeared. Despite its small size it must have been fairly deep, as it never dried out. Snakes and turtles abounded, and for many years a busy covey of quail made the tiny levee home, which they surrendered with surprising nonchalance should a twig-scratched country boy rudely appear. If you approached slowly, and fished silently once you had taken your spot, the birds would return and carry on with their quail work, little ones staying close to their mother in case the strange creature standing nearby posed a threat. What a privilege that was.

We had fished that pond for years with cane poles, with only the occasional bream to show for our efforts. Silhouettes of larger fish out in the middle were a mystery we had not solved; whatever those critters were, they studiously ignored the earthworms and crickets we plunked down as far out from the bank as our homemade poles would reach. But to my knowledge, no one had ever fished that little body of water with a rod and reel. Andy and I were not there that day to catch fish, mind; this was just a training exercise. We worked our way across the levee to the far end where there were fewer trees to hinder casting. Just as Mike had done for me, I patiently explained the workings of the apparatus to my little brother. My demonstration cast, a thing of noble beauty, landed the ridiculous plastic worm in the reedy shallows against the opposite shore. (It was a very small pond). I turned the handle, closing the bail with a metallic click, and gently reeled in the slack line.

"When you feel a fish hit," I said, in my best barefoot instructor voice, "you set the hook, like this." I yanked back, only to have the rod suddenly stop, quivering, beside my ear.

"Dang," I said. "I'm snagged." At which point, my first-ever *big* bass came writhing and thrashing—"dancing," as it is called—out of the water.

"It's not a snag!" I exclaimed, highlighting the obvious. "It's a bass . . . a big ol' bass!" I thought my heart would explode. Out of the corner of my vision I could see Andy's own eyes grow wide as the lunker repeatedly broke water. Bass power increases with size on a logarithmic scale. A half-pounder just skips over the water as you reel it in. A one-pounder puts up a fight. Anything over three pounds feels like a cracked-out ichthyosaur. A determined bass can easily break a line with a listed weight tolerance far exceeding that of the actual fish. And the big ones do not get big by being dumb; they employ a wide variety of tricks, such as flinging the hook while dancing or diving into submerged tree limbs or other cover where they toss the hook when the line gets tangled and goes slack. One compensates for the gut-churning pull of a big one by carefully "playing" the fish and by correctly setting the drag. Being at that time ignorant of such nuances, by rights I never should have landed that bass. I violated every rule in the book getting it in, reeling madly for what seemed like forever with a total lack of finesse against its astonishing strength. As the saying goes, I "horsed her in." Never a good policy. But land it I did, a chunky brute that might have gone four pounds. It looked like forty. That beautiful lunker would remain my PB for a long time to come.

That awesome moment took place fifty years ago, but I remember it like it was yesterday. I still feel the same rush when I hook a big one, and the same sense of breathless triumph when I manage to land it. Such enduring joy, courtesy of a lanky, taciturn, knobbly-kneed companion of former years. I cannot say thank you enough, Mike. Go strut that chicken, my friend.

IV

GEARING UP

Many men go fishing all their lives without knowing
that it is not fish they are after.
—HENRY DAVID THOREAU

The seventies were a curious time to come of age. This was perhaps especially true in Mississippi, a place so endearingly weird that having a nickname like "Taterbug" is a significant advantage should you choose to run for public office. A place so stubbornly resistant to change that Independence Day was not celebrated in Vicksburg, the "Gibraltar of the South," for eighty-one years following the Civil War, while a state flag lacking the Confederate battle emblem first officially tasted the wind only in January 2021. Yet the seventies were very much a time of change, even in our little backwoods county. It was an *in-between* time. You could feel it in the air. You could see it in the cut of women's jeans, a style that, like Art Deco, never has been surpassed in visual effect. Skateboarding became a thing for those who lived where some parts of the earth were conveniently paved. Television was really, really, really bad. We knew it and watched it anyway. Places called *malls* appeared where you could buy all kinds of useless gunk if you had money to spend. (We did not, but the nearest mall was a couple hours' drive away, so it did not really matter.) Guys started sporting long hair, sometimes *made-up* long hair, all wavey and contoured and poster-worthy. Sure, it *looked* sissy, but the girls were going bonkers

31

over Leif Garrett, and the girls were wearing those darn jeans. What was a healthy adolescent country boy to do?

Well, getting a Leif Garrett hairdo was out of the question. For one thing, Pop would have blown his stack at the very suggestion. For another, we could not afford such fancy *coiffures*. Until the ninth grade, Mom cut our hair by setting the clippers to Dalek and buzzing the entire dome down dangerously close to the bone. By the high school years, she allowed for approximately one inch of growth, out of which our ample ears extended like windmill blades. Given everything else that she had to do, one could not blame her for selecting the most efficient method of grooming her young. One blamed her anyway if one were among those so afflicted, especially during those awful years when it feels like you are continuously on center stage in front of a very critical audience. One unfortunate result of the harrowing operation was that, for some considerable time, my nickname in school was—shudder—*Mount Baldy*. Not the kind of moniker to promote confidence in the courting arena. Admittedly, I still would have been awkward and shy around girls had my nickname been Captain Awesome and had I been graced with the male equivalent of Farrah Fawcett curls. Being raised with six brothers and no sisters provides few clues for solving the mysteries of female-kind. Fortunately, I had other things to occupy my attention, including honing my foosball skills and spending endless hours carving the humid Southern air into invisible marshmallow chunks with a homemade pair of wooden nunchuks. But above all other pursuits, there was fishing. I was a willing devotee.

Nowadays, fishing tackle is something I comfortably can afford. In those lean days, not so much. Building up a stash was a serious investment. All of us kids earned money by hauling hay, mowing yards, or doing odd jobs for anyone who would hire us. Some more odd than others. When "city water" finally began to be extended out into the countryside, Bennie dug water line ditches from roadsides to houses, a back-breaking task that he jealously guarded due to the lucrative return of ten cents per foot. His shovel skills came in handy when he and Dennis were hired to excavate chicken crap out of coops that must have seemed to go on forever. Like the fabled Augean Stables, those enclosures were not cleaned with regularity. The poop was two feet deep, with stray three-year-old eggs lying in odious—and odiferous—concealment. As that particular example shows, we never turned down any offer of work, which endeared us to the local clientele who were not exactly rich themselves. I think they also approved of the penal-style haircuts we sported. Our minimum rate was a dollar an hour, unless someone chose to pay us less, in which case our minimum rate was less than a dollar an hour. Regardless, those dollar bills, few though they might be, felt mighty good when the summer sun was shining and the tackle box was growing bare.

One job Andy and I undertook together was using sling blades to clear the roadside bank marking the eastern edge of Mr. Curtis's property in French Camp. "Bank" seems euphemistic; that slope was as steep and forbidding as an Iron Age glacis, and the early successional, sunlight-greedy plants that sprang up between cuttings grew in thick, thorny, and stubborn. It was a brutally hard task that took us a day to complete, a long day during which "them Peacock boys" became the *topic du jour* for the supremely bored worthies loitering around the checkers table outside the downtown gas station. Given that the pay was not exactly commensurate with the labor involved, I felt only the smallest twinge of guilt when, one day during our lunch break, I carried out some covert exploration of Mr. Curtis's old wooden garage and discovered a dust-covered metal tackle box that clearly had not seen use for years. Inside I found—and after an obligatory moment's hesitation liberated—an in-line spinnerbait with an actual feather skirt and a tiny, jointed minnow with a metal lip and two small treble hooks. I eventually pulled those vintage beauties out of commission because they became precious to me. I have both lures still, there in my tackle box with their far more evolved descendants. And if any of Mr. Curtis's descendants should read this account and want them back, I will return the pilfered goods. With regret. But never will I clear that bank again.

A lure put to good effect at the Curtis Road Pond was grub tails, thick plastic things with bodies about the size and color of a cadaver's thumb and having long, thin tails that rippled enticingly through the water. "Mister Twister" was our favorite brand. One variety had *two* rippling tails, an innovation that Andy and I found to be far more irresistible than did the fish. But double-bladed H&Hs, small spinnerbaits with chartreuse rubber skirts, were lethally effective in ponds, rapidly becoming one of my favorite lures. This was a belated discovery given that H&Hs had been manufactured for nearly two decades by that time.[1] They used to be peddled out of a secondhand bread truck;[2] now, you could probably find them in a convenience store in Irkutsk. Not only are they productive, but the sudden, powerful reaction strikes they trigger are far more heart stopping than the genteel bump-bump engulfing of a plastic worm. There are larger spinnerbaits that work very well in lakes and rivers, but for the ponds I love, give me the double-bladed H&H, on which over the years I have caught many, many bass, overly ambitious bluegill, hungry crappie, and a variety of unwelcome reptiles. It is hard to land a really big bass on an H&H, as the hooks are too small and thus easily thrown. But size is not everything, except for The One That Got Away, in which case size is everything.

If there was water on the moon, and were I an hospitable astronaut wanting to do the Earth proud by having something delicious on the grill when the Lunarite neighbors came to visit, and if I could have only one lure because it costs a lot to get stuff to the moon, I would have a Beetle Spin. A small rubber jig beneath a spinning metal blade, a Beetle Spin will trigger strikes from any fish species that ever has lived on the Earth or its sadly dry satellite. There is some dispute over who invented this fish-catching marvel; Chuck Wood, the "First Maestro of Finesse,"[3] or Virgil Ward, the "Sheik," whose favorite high school subjects were "geometry and girls."[4] Like the dispute over calculus between Newton and Leibniz, it really should not matter; humankind was enriched regardless.

I had to have something in which to put my hard-won tackle, which meant acquiring a tackle box. My first one was a small, beige Plano with a fairly capacious main chamber below a single, outswinging tray divided into rectangular compartments. It came with a little metal plate bearing adhesive backing that fit into a recess in the front-center of the lid. I sent the plate and the royal sum of two dollars to the company headquarters in Illinois. I checked the mailbox daily until, several agonizing weeks later, a small, padded envelope appeared. It contained the plate with *my name inscribed in cursive.* I was on top of the world. I took great delight in carefully sorting tackle into the separate compartments of my awesome, autograph edition box. Hooks in one compartment, weights in another; H&Hs in the one larger, square compartment of the tray. The front of the tray boasted a single long, narrow space that would have been perfect for plastic worms except that it did not hold enough, so they usually wound up in the main chamber along with spools of line, corks, stringers, a pocketknife, and various other treasures. I was very proud of that Plano. It made me feel like a real angler.

Tackle boxes were not at that time "worm proof," meaning that one's precious plastic worms would melt into a sticky, gloppy mess in the Mississippi heat, ruining not only the worms themselves but also covering any adjacent items with sticky globs of immutable worm melt. We did not have air conditioning, of course, unless one counts the wagging tails of the umpteen dogs lazing on the porch, so this was an annual problem. Excising that worm gunk was a real pain, although it did provide an opportunity to lovingly inspect, admire, and re-sort all the beautiful lures that had not been overly slimed. It also provided an opportunity to thoroughly clean the tackle box. Tackle boxes accumulate trash: bits of line, empty lure packages, unspeakable biotic remains, lures needing parts that you never actually get around to replacing. And woody detritus. I get a lot of woody detritus, courtesy of pitiable time-management skills. After changing lures, it takes roughly two seconds to work the latch and secure the tackle box lid. Two seconds is an eternity when the fish are biting. Accordingly, I frequently just flip the lid closed, making a mental note to latch it later. Inevitably, I forget the plan, grab the handle of the box, hoist it up, and unceremoniously dump the contents out upon the ground. Or into the bottom of the boat, much to the delight of whoever boat I am in. Once gathered back up all that stuff has to be untangled and resorted, a task usually put off until the evening, especially if they are biting. Hence the transport of woody detritus into the box. It takes approximately three hundred times longer to clean up the mess than it does to latch the box to begin with. No doubt I will dump it again before the year is out.

It did not take long to outgrow the Plano. Over the years, I greedily acquired bigger and bigger tackle boxes in direct proportion to my income, a trend culminating in a monstrous thing the size of a

Soviet apartment block that, when loaded, pulled earthward like airport luggage when the final boarding call is being announced on the far side of the terminal. And man, oh man, when you forgot to latch *that* thing. This was all insane, of course: on most fishing trips, one does not need tackle sufficient to equip the entire Lewis and Clark expedition. Eventually I compromised on a nice, medium-sized unit, the contents of which still would have gotten that intrepid duo at least as far as the Great Plains.

One day, ol' pal Mike took me out in his jon boat to fish Legion Lake, just outside of Winona. The lake lay behind a decaying old skating rink, the favorite hangout of our teenage coterie. Our parents worked hard to persuade themselves that we engaged in innocent and healthy recreational activities there each weekend. In fact, we mostly fastened ourselves to the foosball table, casually showcasing an ability to smoke cigarettes with no hands, or hung out in the parking lot drinking beer, blowing smoke rings, and listening to "You Sexy Thing" by Hot Chocolate. It was all rather innocent, really, if not at all healthy. But teenage invulnerability flowed through our veins, an intoxicant far more potent than the 3.2 beer we consumed in a conscious piece of performance art, dedicated theatricians that we were.

Mike had no boat motor of any kind, so it was all paddle work. But so very worth the effort. For just as he had taught me how to use a rod and reel, on that day Mike taught me the magic of a wondrous lure: the Rapala minnow. Like so many fishing innovations, this sublime artifact has a peculiar history. During the global depression of the early twentieth century, a hard-pressed fisherman named Lauri Rapala spent a lot of time out on Lake Paijanne, the second-largest lake in Finland, where he observed big fish feeding on little fish. Especially wounded little fish that wobbled when they swam. Inspired by this not-so-subtle clue, Lauri used a shoemaker's knife and sandpaper to work a piece of cork into the shape of a minnow. To make it shiny, he wrapped it in metallic foil from a neighbor's cheese packets and chocolate wrappers. To make that shiny covering adhere, he melted down photographic film negatives to create a protective outer coating. What a guy. How he secured hooks, I have not quite figured out. Regardless, the results were spectacular; legend has it that our hero soon was hauling in six hundred pounds of fish a day! His astonishing success led to founding of the Rapala Company, one of the most successful manufacturers of high-quality lures in the world.[5] As the company itself puts it, Rapala is "probably the most well-known Finnish word in international use."[6] Not a bad return from watching big fish eat little fish and asking your neighbor if you could have his chocolate wrappers when he was done.

On that fateful day with Mike, I was trying my usual H&H while he employed his perennial favorite, a "Rooster Tail," basically a cylindrical weight with a treble hook at the business end and a single, willow leaf blade at the other that spins in a flashy circle while the thing is being retrieved. Howard Warden, son of an established big-time lure merchant, made his contribution to history in the late 1940s by developing this striking and impressively effective device. He originally called it the "Retreat Special." One assumes that a troop of horrified admen dutifully sprang into action. Renamed the Rooster Tail, in the 1960s it stormed the bulwarks of the American fishing scene like salmon besieging a spawning ladder.[7] They are among the most beautiful lures in production today.

Storied lures notwithstanding, neither of us was doing any good that day. I was not then the inveterate lure changer that I was to become. Mike already had broken through that wall. "Horseshit," he said, or something to that effect after about his twentieth fruitless cast. He reached into his box and withdrew a small, jointed Rapala minnow. It was a lovely thing, brown on top and silver beneath, with endearing little eyes that gazed about in endless wonder. It *looked* like something a fish would want to eat. I may have made a few chomping motions myself. Mike made a practiced cast toward the shoreline, landing the lure between the branches of a partially submerged fallen tree. Rapalas float until you start reeling. Mike just let his float. The little impact ripples receded. Time stood still. He twitched the little minnow. And twitched again.

Ba-*loosh*!!

Ask anglers what is the best kind of strike, and a wistful glow will light in their eyes. "Top water," they will say in the same reverent tone that small children use to entreat with a mall Santa. "Top water." Fisherman drool is not the prettiest sight to see, but it cannot be helped. The stimulus is too compelling. Small bass make a "*plork!*" sound when they hit a topwater lure. Decent-sized bass go, "ba-*loosh!*" Big ones go, "ah-*OOM!*" like a cannonball hitting the water. Or so it sounds to eager ears. As a young boy, I forced myself to stay awake while watching the first moon landing on a big black-and-white TV in our living room. Took Armstrong *forever* to come out of that lander. That is what the intervals between *ba-looshes* and *ah-ooms* feel like. But like Christmas morning, the result is worth the wait. Find the fish on the right day and topwater lures are your ticket to glory. That day was right. As Mike began boating bass after bass, I humbly asked if he had another minnow in his box. He did. We wound up with a real mess of fish that day, and I never will forget the excitement of those first topwater strikes. My tackle box has not been without Rapalas since.

That experience broadened my tackle mania to include topwater lures. Many lawns were mowed and much hay was hauled as I dis-covered the many kinds that were available. Their development is the stuff of legend. James Heddon fished a small mill pond outside of Dowagiac, Michigan, in the late 1800s. According to a somewhat apocryphal story, the good James one day was killing time whittling by the pond while waiting for a friend to show up. Upon arrival of said friend, he tossed his piece of whittled wood out into the water, triggering a bass blowup.[8] James knew a good thing when he saw it. A multitude of Heddon fishing marvels soon followed, including the devilish little Tiny Torpedo, a colorful plug with attention-grabbing eye spots and a fetching little propellor blade comically positioned at the butt. To compete with Heddon, in 1938 Fred Arbogast intro-duced the Jitterbug,[9] a stubby, cigar-shaped plug with a metal lip like a snow pan that goes "*bloop*" when you twitch it, followed soon after by the Hula Popper, a sort of frog-shaped thing with a hula skirt and a scooped mouth that goes "*blorp*" when twitched.[10] "*Bloop*" and "*blorp*" are bass-speak for "Annihilate me, please." Drool.

A variation on the Heddon story has him whittling out a wood-en frog while waiting for his friend to show up, using a broomstick handle for raw material.[11] Frogs are extraordinarily effective bass bait, so much so that a very early, and very horrible, contraption, "Ketchum's Frog Casting Frame Gang," was patented in 1904. It was like a medieval rack upon which actual, live frogs could be emplaced, an appalling contrivance that "Never Fails to Ketchum," as the box proudly proclaimed.[12] Fortunately, artificial frogs already were available and soon became a standard lure. The first rubber frog was the Hastings Weedless Casting Frog, which hit production in 1895![13] An enormously varied series of frog lures soon followed, all of more-or-less cartoonish appearance, contributing to their status as among the most collectible of lures.[14] Comical they may appear, but nothing entices a big bass quite like a frog. They are hard to fish, as one must keep one's reflexes in check for long seconds until the bait is fully sucked in. I have never quite gotten the knack, but I cannot resist throwing frogs anyway. They are too much fun not to.

All of these topwater lures can produce wonderous results. But buzz baits . . .! A buzz bait is like a spinnerbait, except that instead of a blade at the top of the wire "V" that frames the lure, there is a multi-bladed propellor. This propellor keeps the lure at the top of the water during retrieval while making a most satisfying chopping sound as it cuts the surface. Especially on overcast autumn days, bass will do their damnedest to take out a buzz bait. It just makes 'em mad. Around in one form or another since the forties, buzz baits really took off with popularization of big-time fishing tournaments in the seventies,[15] for the simple reason that really big ones will hit a buzz bait. And they do not just hit it; they *smite* it. That is a heart-stopping moment, especially given the possibility that one might just have grasped the Holy Grail of Bass Fishing: the double-digit bass. Most anglers never hook, and still fewer manage to land, a bass weighing ten pounds or more. A gentleman in Valdosta, Georgia, named Pat Cullen has caught well over *one thousand* double-digit monsters using a large, black buzz bait at night.[16] His nervous system must look like a circuit board.

The kind of buzz bait I discovered does not, alas, seem to be in production anymore. It was a small lure, with a greenish-yellow, tri-finned propellor riding over a medium-sized hook hidden within a chartreuse skirt. It made a lovely sound and could be retrieved very slowly, maximizing the potential for strikes. Mr. Henning, father of some school chums, owned a small pond out in a pasture near our house that he graciously gave me permission to fish. That pond had a healthy component of lily pads interspersed with tempting little open spots of water. The buzz baits were so light that you could drag them right over the pads, only for them to start chopping the water again as soon as they hit an opening, where, as often as not, a chunky two- or three-pounder would go off like a depth charge. The hook sets were glorious, and fighting to get those fish to shore through the intervening lily pads added a thrill of terror to the endeavor. I had found my first honey hole! Eventually, Mr. Henning not so graciously suggested that I leave some fish in his pond. It was a hard request to honor. But it was a valuable lesson: *never* outstay your welcome at a prime fishing spot.

Hopelessly enamored with topwater strikes, I saved up and bought a fly rod, enormous fun in the South, where bluegills attain the size of personal-pan pizzas. But the fun really started when I discovered popping bugs for bass. These are basically Jurassic-sized versions of the little popping bugs used for bream, colorful little plugs with feathered tails and/or rubber whiskers. When a bream hits a little popping bug, it goes, "*blip!*" and the fight is on. For its size, nothing fights quite like a big ol' bluegill. But when a bass hits a big popping bug, there is a mighty "*thorp!*" followed by a jolt when the hook sets after flinging back the long, flexible rod. The line is pulled in by hand, angry bass fighting all the way. It is quite the sensation. Norman Maclean suggested that watching a true fly fisherman at work is very like a religious experience. I once watched my friend Paul, a lifelong fly fisherman, hook, play, and land a huge carp. That was indeed something like divine. Alas, the faithful would have fled the temple in droves after watching my untutored technique. But I had a heavenly time. It was so much fun that I took to wading out in the more grown-up ponds to access tempting places along the edges. Eventually I began to realize that providing a fleshy landing platform for undulating moccasins probably was not the best idea ever. I purchased some chest waders, which at least provided a false sense of security *in re* the reptile factor. The voluminous waders had other uses as well. One Christmas Eve, I snuck into the living room late at night and used them to replace my cloth stocking. Santa was not amused.

Other than Mike, who was self-taught, I had no fishing instructors, although I watched Bill Dance fall out of the boat on TV every chance I got. Experience thus became my teacher, at least until 1980 when *Roland Martin's 101 Bass-Catching Secrets* came out. That was where

I first learned about *patterns*, combinations of environmental factors and tactics that can produce remarkable results. It was my first lesson in really strategizing about fishing, and I still employ tactics learned from that book. But there are things undreamt of in even the pros' philosophies. One gaspingly hot summer day circa 1978, Cousin Ted and I hit a typical farm pond, a small, square pool in the middle of a bleached and desolate pasture. Raised a military brat, Ted embraced country life with admirable gusto, especially where sharp things and gunpowder were involved. But at that time there was much he did not know about rural mores. He may even have pronounced crappie as though referring to a bodily function. In relative terms, that made me a seasoned veteran. Loftily, I invited him to help himself to my tackle whilst I whizzed an H&H out over the water. H&Hs seldom fail me. On that day, I might as well have been throwing shade.

"How about this?" Ted asked. I turned to look. My jaw dropped. He had threaded at least a dozen plastic beads of different sizes and colors onto his line, below which he had not one, not two, but *three* lead worm helmets riding atop a shiny, golden hook. Onto that hook he had threaded not an entire plastic worm, but three or four pudgy *chunks* of plastic worm. It looked like a piece of intestinal tract from Mechagodzilla. Through the tilted lip of disdain, I confidently pronounced that nothing, *nothing*, was going to bite that silly contraption. Undeterred, Ted cast out . . . and, of course, immediately hooked something. That something was long, lean, and toothy, with a beautiful color pattern and malevolent eyes. It turned out to be a chain pickerel, of all things. Not that I knew it at the time, having never before beheld such a fish. How pickerel got into that little farm pond, I cannot say. But to my chagrin, Ted kept pulling them out one after another. Humble pie is lean fare, but I ate my share that day as I replicated his ridiculous lure and caught a few pickerel myself. Another valuable lesson: in fishing, as in the rest of life, things do not always go by the book.

It was about that same time when I had my first encounter with a true monster of the deep, the alligator gar. Cousin Roy lived in Cleveland, Mississippi, a pleasant little Delta town where the wonders of civilization abounded, including a bookstore and a public swimming pool. I shamelessly invited myself to spend time there whenever I could. Roy was, by our standards, exceedingly wealthy, one indication being his actual, real, honest-to-goodness, sure-nuff *boat*, a beautiful thing with a huge motor hanging off the transom and a helm protected by a windshield. It had a steering wheel! To those of us who were boatless and who spent far too much of our formative years on Gilligan's Island, it truly was a marvel to behold.

One fine day Roy took Andy, me, and his son Brian out onto the Mississippi River in that stately craft, about as heady an adventure as could be imagined. The high point of the day was anchoring above a sunken barge and dropping crickets down on lines buoyed by porcupine quill floats, long, thin, brown-and-white striped magic wands that slid enticingly down into the water as bream after enormous bream devoured the bait, completing the spell. The low point came after we tied onto a semi-submerged houseboat next to the bank. We all clambered onto the thing and fished over the railing, which was great fun, especially when a torpedo hit my line. The fun ended when I managed to land the torpedo. The gar that I boarded with great difficulty was a small one; they can grow up to eight feet long and top three hundred pounds. Mine might have gone three pounds, but the dragon scales, the snake-like body, the gimlet eyes, and especially the double row of needle sharp teeth were terrifying, nonetheless. The scales looked like overlapping arrowheads, a function I later learned they actually served for Native Americans in pre-European times.[17] Roy watched with great amusement as I tried to extract the hook without losing a finger. He was even more amused when, a bit later, another gar slammed my hook and tied the line up on a submerged tree limb. At his insistence, I had to go *into the prehistoric-monster-infested water* to free the thing. In retrospect, I think he was just getting revenge for my having caught more bream than he.

In addition to lures, I became entranced by rods and reels. For years, an ad ran on TV for the Ronco "Pocket Fisherman," a little plastic device that still is made today, although with much sleeker lines. The version I coveted looked like something Dr. McCoy might have used for proctological exams. It had a brown, oblong body and a funny little split rod bearing exactly two eyelets. I know people caught fish with those things—I saw them do so on the commercial, at any rate—but how you could land anything over a pound, I could not imagine. I badly wanted one. Another innovation, cunningly advertised on the back inside covers of true crime magazines, comic books, and other such edifying media, was the "Amazing Autocast," the "Fishing Rod That Shoots Lures Like a Gun!" I never owned that device, either, which is just as well, as Hardy doubtless would have put it to nefarious use. Which might have led to an interesting piece in *True Detective*, now that I think about it.

My first encounter with a rod and reel other than the Zebco was with another closed-face reel, an elegant little Daiwa Minicast with a four-and-a-half-foot, light-action rod. Robert had received it as a birthday present. Jealousy reigned. He did not use it *nearly* enough. One day while Robert was away, I surreptitiously borrowed his birthday present. I walked down the gravel road, past an old house place marked by an enormous, fallen white oak and a solitary, hand-levered water pump, to a little pond surrounded by hardwoods on three sides with a grassy, south-facing levee. The pond belonged to our nearest neighbor, Mr. Atkins, who once swore that he'd seen a UFO settle to the bottom. For years I had sat upon that levee with a cane pole, determining beyond doubt that earthworms are not considered desirable fare by interstellar tourists. Never caught an alien aquanaut, anyway. Instead, I caught ridiculous numbers of voracious little green sunfish and tiny bluegill, yanking them out as fast as I could get a hook into the water. On that day, though, I was there for another species altogether.

I caught a grasshopper and stuck it onto a small hook, lowering that into the water without even bothering to attach a cork or weight. Even the grasshopper was somewhat superfluous, as I was after green sunfish, the minute piranhas of the freshwater world. Green sunfish will eat anything. I have caught many on a bare hook. Lowering the grasshopper into the water was like the scene in *Jurassic Park* where they lowered a cow into the velociraptor cage. It took no time at all to pull out three of the little blighters. I tossed them onto the bank and rerigged, tying a big, rusty lug nut onto the end of the line, then coming up about four inches and loosely pinching the line together to form a loop that I pushed through the eye of a pretty big treble hook. After securing the hook in place with something resembling a fisherman's knot, I was ready to tightline for catfish.

I used my Barlow pocketknife to cut up the sunfish, an act of mercy given that they already had been discovered by fire ants. I then moved down the levee a bit as I, too, had been discovered. Karma, I suppose. I impaled a head section with the hook and whizzed the horrid thing out over the water. Once it hit bottom, I slowly tightened the line until the tip of the rod was slightly bent. Then I waited, sitting cross-legged, handle in hand, while the Mississippi sun baked my already browned skin. Karma morphed into Zen. I had not waited long before there came a gentle bump, bump: a pause, and then a strong, steady pull. I dropped the rod tip, reeled up the slack, and set the hook. The action was terrific as I played and landed a feisty catfish. Several others soon followed. What a sweet little rig that Daiwa was! The balance was perfect, and the feel was like nothing I had experienced before. I might even have asked permission next time I used it.

Now that I have reached a stage in life where I can afford far more gear than I should buy, I still occasionally feel nostalgic for those days when a grub tail was a serious investment, few pond fish had yet seen an H&H, and a rocket-powered autocaster promised a consummation devoutly to be wished. Plus, there were those jeans. As in-between times go, it could have been a lot worse. I might have taken up interpretative candle making.

V

CASTING INTO THE WIND

Give a man a fish and he will eat for a day. Teach him how to fish,
and he will sit in a boat and drink beer all day.
—GEORGE CARLIN

Fishing can be a liquid sport, especially in the Deep South where there is a strong cultural link between alcohol and pretty much anything having to do with the outdoors. Always more pious than practicing, Mississippi was the first state to ratify, and the last to repeal, Prohibition. Drinking nevertheless abounded in "The Wettest Dry State."[1] The recipe for a drink called "Mississippi Punch" first appeared in 1862,[2] Angostura bitters reflecting, perhaps, the sour tang of secession. During the bleak years of the early twentieth century, it was the enviable man indeed who had access to both a whiskey still and a fish weir. Traps were set for paddlefish and revenuers with equal determination. It was a hard, hard time, out of which came people who drank life like moonshine and vice versa. People like my mom's dad.

Granddaddy had his ways. He was a lovable rogue with a face like a cartoon president and a chortling laugh that often broke into spasmodic coughs, courtesy of the enormous cigars he could not do without. Nor could he do without nasal spray, which he spritzed up his ample, cratered nose many times daily, bringing the nurturing droplets closer to his brain by lusty in-draws of air through the very nostrils he was trying to clear. The sound was unspeakable. Granddaddy habitually wore coveralls, some the same unnatural reddish-orange color as the top of cheap Styrofoam fishing floats. He slept under an electric blanket with the air conditioning set to Late Pleistocene. He had chests of drawers stuffed full of socks. I mean, *lots* of socks. His latter career was as a night watchman at Parchman Penitentiary, an ironic posting given the liquor consumption, dice games, blithe philandering, and cheerfully executed petty larceny, sometimes involving his grandkids as willing accomplices, that characterized most of his adult life. I suspect that the coveralls came at a remarkable discount, unbeknownst to his employer. Granddaddy glided across the landscape in his enormous Buick, drifting into this lane or that while chortling himself into yet another coughing spell, cigar and steering wheel in one hand while the other held a handkerchief into which he expectorated life. How we loved that man.

Granddaddy was one of the most avid, and perennially jolly, fishermen I have ever known. He loved nothing better than to be sitting on a riverbank tempting the crappie with live minnows, especially if there was a cooler full of beer at hand. He lived his later years in Tillatoba, a tiny settlement conveniently located between Enid and Grenada lakes. Grenada especially was a favored fishing spot, a sizable Army Corps of Engineers impoundment centered on an impressive floodgate barring the Yalobusha River on its way to the broad, muddy Yazoo. The spillway was a magical spot. Wooden picnic tables bearing the yearning inscriptions of many a nominally literate teenage swain cooled in the shade of enormous pine trees. Big, green metal barrels bore signs cordially inviting people to deposit their trash within. Styrofoam cups, empty lure packages, soda bottles, and cigarette butts were the surrounding archaeological landscape of a confounding species. Especially when they lowered the reservoir after rain, water came shooting out of the spillway, thundering down a broad concrete ramp to churn and mix in the channel below like a roaring mountain stream. Mississippi not being a place of roaring mountain streams, it was a marvelous spectacle for a kid to behold. The air was saturated with heavy hanging droplets, especially early in the morning when fog cloaked the scene in mystery. One could get chilly there. In summertime! The whole place smelled strongly of fish, which was not surprising given the innumerable gar skeletons in tattered dragon-skin shrouds decorating the blue-gray riprap lining the banks. Getting to the water was a dangerous enterprise, as the rocks could shift unexpectedly beneath one's feet. Lost hooks, weights, and bobbers were free for the taking between entrapping boulders. It was like reaching into Aladdin's Cave.

On rare and glorious occasions, Mom, Pop, and all us kids would join Granddaddy to fish the swirling, lacquer-colored water, reaching out from the bank with homemade cane poles, repeatedly lifting and dropping the line after the current quickly swept the cork downstream. Meanwhile, devotees with equipment worth more than our truck cast into the deep pool at the base of the spillway, targeting the enormous catfish known to lurk there. A popular story was that a diver once went down to inspect something about the dam and, upon surfacing, announced in shaking tones that he would not be going down again lest he be devoured. I since have heard the same story at many other spillways, which hardly diminishes its charm. But big ones, there certainly were: we once watched a guy haul out what must have been at least a thirty-pounder. Someone told us it was his second big one of the day. Those whiskered behemoths must have tasted like long-submerged crime-scene cars. But oh, to land one! One such trip to Grenada involved an overnight stay at a roadside inn. Whenever Granddaddy and Pop got together, all token restrictions on liquid consumption were off. Recovering on the morning after, Pop tied his fishing line to his toe and lay back for a restorative nap in the sun while his cork yanked and plunged in the eddies by the rocks. He caught more fish than I that day, the rogue. Fortunately, he did not get yanked in by a passing crime-scene car; swimming in such waters would have been incredibly dangerous.

Granddaddy had a standard jon boat with a vintage outboard motor. His view from the back bench frequently was of the blue Mississippi sky, not out of appreciation for the aesthetics of the situation but because that is one's perspective while one is draining a can of Schlitz while carefreely motoring at high speed over the watery deep. But not that deep. Encounters between the outboard propellor and submerged stumps were frequent, such that Granddaddy kept a dozen or so extra shear pins in his tackle box. He was amazingly adept at replacing the things, which was not surprising given that on a typical fishing trip he might change them out two or three times, popping another top and revving up to high gear once the operation was complete. While boating, his customary cigar was replaced with Alpine menthol cigarettes, perhaps because there were fewer sparks to set his hair on fire as he ploughed into the wind. That man got a lot of joy out of life.

And that is just how it was in that place, at that time. The men drank. And hunted, and fished, and watched football on TV, voices growing loud. The women drank too, of course, most of them. But the men seemed far more determined to call attention to the fact. We young ones watched and learned. Margaret Mead once said that if you want to know how to operate successfully in another culture, do as the children do. Were you observing the rites of passage for boys in our neck of the woods, that would mean feeling the kick of a twelve-gauge shotgun and the kick of a twelve-alarm hangover by the age of sixteen. Some of us started earlier than that.

It was not just near kin, by any means. A sixth-cousinish branch of the family occupies a substantial portion of the hills around Stewart, Mississippi. Like so many "census-designated places" in the Upland South, Stewart's glory days were in the early twentieth century, when timber was being transferred from Mississippi to booming northern metropolises as fast as locomotives could carry it. Once the state had been essentially clear-cut, such homely little places went into decline, like aging relatives forlornly marking time in a nursing home. You visit when you can, but slow loss is the hardest to bear. By the time I was old enough to drive, about the only things left of Stewart central were a sublime old wood-framed general store and a totally incongruous, postage stamp–sized washateria that looked like it had been glued to the landscape by some unsteady hand. I used to help Mom haul basketloads of wet clothes to that washateria when weather prevented her hanging them out at home. The building was dismal, inside and out, one of those places that somehow manages to be at once sterile feeling and trashy looking. And smelling. The funk leaking out from around the battered dryer doors was like a mildewed funeral shroud. Those washateria visits were not my idea of a good time. But navigating the series of shaded, gravel roads between the two locales was fun.

Knowing those roads was laying claim to a shared identity. We might have been hill-folk, but they were our hills, dammit; our woods; our roads. Our fishing holes! It was okay that we might occasionally pilfer from one another, as long as the take was small and a blind eye was turned when reciprocity came calling, but woe betide the interloper from Jackson caught field dressing a deer or casting a line into someone's pond. We teens marked our territory with Miller High Life bottles flung from open windows as we careened around curves, spraying gravel as we purposefully fishtailed the rear-wheel drive vehicles that were all anyone had back then. Fortunately, traffic was light. A car passing by our house was an event to be noted with wonder. Or suspicion.

Paul and Nathan were of the Stewart clan and of my generation. We were fast friends, school chums, and lords of all we purveyed out there in the sticks. We shared that domain with others of our bent, a gang of aspiring delinquents whose friendships were close bound enough, as Graves wrote. Our territory encompassed a rough triangle between French Camp, Stewart, and Winona, familiar settlements connected by a venous system of tertiary roads down which sediment flowed like plasma following rains. The blacktops were faster, but the journey was the purpose, really, and there were significant advantages to taking the long way around. You never met a cop out there on those country byways.

It was on a fishing expedition with the Stewart clan that Paul, Nathan, and I helped gather large numbers of spotted salamanders from beneath logs in the Big Black River bottomlands. Nowadays I would not murder a salamander to save my life. In those days, it was all part of a culture where one could boast about one's respect for the outdoors while routinely committing multitudinous crimes against nature. The salamanders were used to bait large hooks secured by thick, white braided line to short, stout cane poles shoved into the banks of the river. We ran the river in jon boats early in the morning for a couple of days, catching scores of large catfish, their predicament signified by the pole yanking down toward the water, a sight to light a fire in the heart of any fisherman. Heaving those weighty monsters up out of the muddy deep and into the boat was a gratifying experience. Watching a grown man (of no relation whatsoever, please note) clamber out of the boat onto a sand bar, grab a writhing fish and split it open, then scoop out the roe and devour them raw while saying "Caviar!" in whiskey-soaked tones, was not.

A more hair-raising adventure with some of those same shining exemplars of adulthood was "frog grabbing" on the Big Black, an enterprise that caught the fancy of my imagination only to raise the specter of imminent death once underway. There were three of us in a fourteen-foot jon boat. As the neophyte, I occupied the central bench, my feet in what I soon discovered would be the dead well for dozens of large, slimy, freshly brained bullfrogs. The grabber was stationed in the front, half-kneeling on the metal bow plate, holding a spotlight in one hand and a can of beer in the other. The pilot sat on the rear bench, handle of a twenty-five horsepower outboard motor in one hand, can of beer in the other. It was after dark when we launched, a moonless night that made the Big Black bottomlands especially creepy. I have been lost in those bottomlands. It was not an experience I care to repeat. Neither is frog grabbing. We tore off down the river, the pilot obviously unable to see any floating, or worse, semi-submerged obstacles in front of us given the supercargo obscuring his view. How he was navigating, I have no idea. He could have piloted by searchlight except that Queequeg was using it to scan the shoreline as we shot past. Abruptly, Queequeg yelled "Frog!" whereupon Ahab shoved the motor handle hard over, flinging us sideways and casting a very impressive starboard wave as we shuddered to a near-stop in the river. On the bank, in the glow of the spotlight that Queequeg held remarkably steady in one hand while somehow holding on to both his beer and the rocking boat with the other, was a gray shadow on top of which two ridiculous, googly eyes shone brightly in the night. Anyone old enough to have watched the original Looney Toons would have known immediately what it was we were looking at; the resemblance to Michigan J. Frog was remarkable. The boat now more-or-less perpendicular to the shoreline, Ahab opened her up, full throttle. As we slammed against the bank, Queequeg snatched up the mesmerized frog, falling backward into the boat with the rebound. Even as he fell, he was repeatedly slapping the frog's head against opposing gunwales, transcribing a grisly arc in the air with the grotesquely extended amphibian.

"That's one!" he said, dropping the quivering frog at my feet and reaching for the fallen beer can spasmodically vomiting cheap lager into the boat. And onto my shoes. Ahab cackled as he hit hard reverse, the outboard laboring as we slowly slid from a rectangular divot in the muddy bank. That improbable scene replayed many times that long night as we ramped at high speed over logs and stumps, swinging hard to ram the bank whenever another googly-eyed silhouette appeared. I graciously declined an invitation to try my hand at grabbing, as the floodlight sometimes illuminated impossibly huge moccasins probably hunting the same frogs we were. Paul and Nathan were in another boat somewhere else on the river. At some point I realized that, if we lived, we were going to have a lot of frogs to clean. At least there were no age restrictions on alcohol consumption. Such were our role models of the time.

Not surprisingly, we grew up more than a little rough around the edges. One spring day, ol' pal Mike and I made our way to a lake deep in the woods behind the skating rink, banging along across the Legion Lake levee and down a mostly nonexistent forest track in his little red Toyota pickup. Somewhere along the way we made the impromptu decision to camp out. We had brought the basics, after all: fishing rods, tackle boxes, cigarettes, matches, a .22 rifle, and beer. What we had not brought would have filled a much longer list: tents, sleeping bags, pillows, blankets, cookstove, cookware, utensils, food, water, flashlights, lanterns, jackets, insect spray, or any number of other such namby-pamby accessories. That was not the way we rolled in north Mississippi in the seventies.

We had a fine old time that day, even though the bass were not biting. As twilight fell, the mosquitoes were. We started a fire to fend them off, then stuck the handles of our rods into the squishy bank and set up tight lines, hooks baited with chunks of bluegill caught from the water's edge at dusk compliments of the ever trusty Beetle Spin. Sitting by the fire, drinking beer, shooting the breeze, watching the stars come out, waiting for the poles to dip; that was living. As the hours passed, we hauled in a healthy number of catfish, dutifully lacing them to a stringer that we stuck to a firm spot in the bank far enough away to avoid entanglement when more fish were brought in. Sometime around midnight, Mike said, "What's all that damn ruckus?" or something to that effect. A loud splashing was coming from the direction of the stringer. About the time I realized that a large moccasin was taking hungry bites out of the captive fish, a rifle spat very close to my ear as Mike put a bullet into the snake's head with nothing more than flickering firelight to illuminate the scene. That was not luck. I once saw him similarly shoot a swimming snake in the head, at a considerable distance, while standing up in his wobbly little jon boat on a windy day. On both occasions, he needed only one shot. We grew 'em tough in the Hills, if not particularly conservation minded.

Not long after the snake incident, the deadwood we had gathered for fuel ran out, as did the beer. The want of any sleeping gear suddenly became more than a minor inconvenience, as the night had grown damned cold by that time. Lacking other options and displaying an unaccustomed wisdom in choosing not to drive, we simply lay down upon the ground beside the dwindling campfire and shivered our way into a few hours of fitful sleep. When I climbed to my feet the next morning, dew soaked and aching, I knew for the first time what it would feel like to be old. Our snake-chewed fish were ruined, hunger was gnawing at our empty insides, and it was many miles back to Mike's house and a desperately needed pot of coffee. That was an expedition typical of the time. I felt very lucky to have been there.

Three days out of high school, just seventeen years old, I joined the US Air Force. For most of us Airmen, it was our first brush with adulthood. Pretty sure it was a wire brush. But it had its moments, especially the frothy ones. The local lager suppliers were happy to see us coming wherever we were stationed. We were happy to keep them in business. Contemplations about the future were effectively submerged. We worked, and we played, and we drank, and we talked. We talked about girls. We talked about music. We talked about movies. We talked about TV. We talked nonsense. And we talked fishing. That was both infinitely fascinating and sadly desperate when I spent two years stationed in Germany, homesick not least because I was unable to wet a line. The warm air of spring spawning season created a literal ache in the bones. Enviously, I watched from train windows as well-dressed Deutsche used ridiculously long poles to lift small, silvery panfish out of broad, placid canals. I badly wanted to join them, but getting a license there was an expensive and complicated process, especially for a foreigner. In their benevolence, the Fishing Gods decreed that my next duty station be Eglin Air Force

Base, a government installation taking up an enormous chunk of the Florida Panhandle. I had scarcely tossed my duffel bag into the barracks before I bought a sweet little Mitchell reel paired with an exquisitely balanced, medium-action rod, and some basic tackle. I had a lot of catching up to do. That Garcia Mitchell 218 was my first open-faced spinning reel. It was a beautiful thing. Spinning reels are gratifying devices because they are so tactile. You hold the line across an index finger and flip the wire bail over to free the spool. You feel the weight of the lure pulling against your finger, and an experienced angler will let the rod tip drop until the perfect balance is achieved, data streaming into little gray cells that might otherwise be usefully employed. Data conveying how hard to cast, and about how far the lure is going to go. Casting with a spinning reel is a sublime experience, except for the all-too-frequent cases where excitement fuses the frontal lobes so that data related to tree limb avoidance arrive tardily. If the circuits are functioning properly, you let go of the line at just the right moment to achieve a sub-orbital trajectory, landing the lure with William Tell precision in the exact spot that your fishing partner has his eyes on as he madly reels in his own lure to try to beat you to it. As it happens, circuits fail with surprising frequency; even the experts get snagged, especially in the competition to hit the best spots. A major advantage in that competition is to be in the front of the boat. How is that for a metaphor for life?

The Mitchell was invented in France by Maurice Jacquemin, chief engineer for clockmaking firm Carpano & Pons, who contracted with a tackle company called La Canne à Pêche ("The Fishing Rod") to develop a reel. Jacquemin's mighty brain was housed in a head shaped like a mushroom cloud, and the results of his efforts were fittingly explosive. His revolutionary design led to booming sales, especially when the Gumprich brothers, Jules and Otto, and Otto's marketing genius Tom Lenk, used their Garcia & Company firm to crack the fishing market in America. Their efforts were greatly aided by all the good ol' boys returning home from Europe after World War II. Along with mementos from overseas sweethearts and the odd battlefield souvenir, many GIs brought back Mitchell reels. But for a French law forbidding the use of proper names for products, the reels might have been called "Michels," after Jacquemin's son. One suspects the Anglicized version of the name helped with sales due to its more American feel.[3]

Fishing in Florida was very different from what I was used to. The water in the ponds and lakes was startlingly clear, requiring a kind of predatory finesse that I had not developed probing the turbid waters of Mississippi farm ponds. There were other epiphanies as well. At first, I followed my standard practice of wading out up to my sternum in order to target the best spots along the banks. One day I clambered out of a large, clear lake to find myself standing in a roughly circular area of oddly flattened grass. A mound of reedy detritus in the center looked sort of like a small beaver house. I was, I suddenly realized, standing in an alligator nest. Alligators in Florida can top fourteen feet in length and eight hundred pounds in weight. I was five-seven and about one-sixty. That was the last time I waded in Florida.

Eventually, I moved out of the barracks and into a crappy little trailer that I shared with Andy, a Midwesterner whose appetite for liquid delinquency may have exceeded my own, although I was not ready to yield my title without a fight. Andy and I tightlined many a night on the banks of streams feeding Choctawhatchee Bay, drinking Canadian Mist whiskey out of a shared bottle while waiting for the sudden, urgent pull of a catfish strike. The fish followed a bidaily cycle, moving up the creeks when the tide was high then back down to the mouth at low tide. Those were big fish. Their passing was signaled by our poles suddenly bending toward the creek, a moment when the bottle was dropped as we lunged to grab the handles before our rods were pulled free from the sandy bank. If we missed the strikes, we got skunked, but it did not matter. What mattered was being there. Andy was a great guy. In the wee hours one morning after we closed the bars, he suddenly decided that the next order of business was to go fish a nearby pond. All hands were, I assured him, on deck, wobbly sea legs notwithstanding. We made it to the pond as the sun was rising. Andy later told me that all he could remember about that morning was seeing me repeatedly fling my lure out a good foot and a half from my cross-legged position on the bank, a position that at least kept me upright. All I can remember was seeing Andy lose his balance and stagger out into the water up to his waist, a memory that

I treasure to this day. I say that is all I can remember; I also remember that I was using a Beetle Spin.

Steve was another fishing buddy at Eglin. Somehow, he found out about a mothballed Coast Guard base on the central Florida Gulf Coast that military personnel could access for R and R. The western edge of the base was pristine shoreline, off limits to civilians for who knows how many years. It was an ideal fishing camp, and we had the whole place to ourselves. Naively, we had brought along our freshwater rigs, being ignorant as to the dire effects of salt water on reels. But ignorance is bliss, at least in the moment. And what a moment it was: blue sky, beautiful beach, bright, rolling water, and not another soul in sight. As we were baiting up for our first try, a pod of dolphins came cavorting by, not twenty yards away. It was a marvelous, magical day for a boy from the Mississippi hill country to be alive, young, and fishing.

Neither of us had any idea about how to properly surf fish, but gamely we cast thawed bait shrimp out into the waves. I got my line into the water before Steve, much to his chagrin. Even more to his chagrin, I immediately caught a fish, a long, solid, silvery thing with a pronounced dorsal fin and a doofy-looking mouth. I tossed it into a small pool on the sandy beach and baited up again. Again, an instant strike. In short order the pool began to fill up with our catch. Whatever those things were, they fought! It was awesome. With typical dumb luck, I won the day. Awesomer still. When we had a few dozen fish flapping around in the little pool, we chunked them into a cooler and rode into the nearest town to: a) score some more brewskis; b) find out what the heck it was we had been catching; and c) determine whether they were edible. They turned out to be whiting, which, the locals assured us as we bought their hideously overpriced convenience store beer, were toothsome in the extreme. Steve cooked up a bunch that night. Toothsome in the extreme, they were not. Relative to Elkins Creek bream, they were, in fact, rather gross. Could have been Steve's cooking, but feeling magnanimous following the day's victory, I suggested no such thing. Plus, I paid for the beer. Nature finds a balance.

After my enlistment expired, Mississippi and a series of desperately unappealing jobs awaited, culminating in a dreadful period nailing asphalt shingles to blazing hot rooftops for much pain and little pay. At least I had my fishing gear, and ol' pal Mike was still around, so there was some escape from the dreariness of the workdays and the emptiness of evenings spent at grimy beer joints. One expedition was a return trip to the lake where a snake had eaten our fish years before. This time, we had Mike's battered old jon boat, and we made a slow circuit around the shoreline. Still no motor, of course, but Mike was one hell of a sculler. As we neared the rip-rapped levee, crappie started slamming our Beetle Spins. At the time, I had so little experience with the species that I did not know we were catching absolute toads, many of which would have gone two pounds or more. We just hauled 'em in, strong fighters and fine eatin' that they were. The rarest of Mike's rare laughs was a single, heavily aspirated chortle that burst forth whenever life surprised him with an unexpected gift. That laugh was rare for a reason. It always was accompanied by a delighted grin, an equally rare gift that I received more than once that day as we just slap wore 'em out. Artists have created acrylic cubes in which, preserved against time in startling clarity, are the most delicate and beautiful flowers. That day on the lake with Mike was like that. Had I said such a thing at the time, he would have shoved me out of the boat, of course. But he would have come in after me if I needed him to. That is the kind of friend Mike was.

It is hard to pinpoint the time when I realized that drinking was not adding anything to such experiences. Rather, it was a distraction from the things that really mattered: paying attention to the environment, playing the fish instead of horsing them in, sharing those timeless moments with good fishing buddies. Plus, it pays to be steady on one's feet while peeing from a boat. Fishing needs no mixers to be intoxicating. I am always ready for another round.

OVERBOARD

Fishing is unquestionably a form of madness,
but, happily, for the once-bitten there is no cure.
—LORD HOME

College life brought many things, vastly broadened horizons not least among them. Most importantly—and most unexpectedly—it brought a loving partner and two precocious stepchildren. Time to grow up.

Janet and I were an unlikely couple. Some fourteen and a half years older than I, she was an accomplished scholar, a seriously invested teacher, an indefatigable worker, and a dedicated parent, not to mention truer to her principles than anyone I have ever known. I tried to emulate her on all such fronts, but she set a mighty high bar. I became a better person due to the effort, though. For a while, she humored me by coming fishing, and we even bought the kids some gear. David astonished me one day on a private lake by reeling in crappie with a baitcaster as though they were going out of style, while Nikki heaved big bass up from around a dock, laughing gleefully all the while, and Janet helped haul in catfish on a trotline. That was a good day. Mostly, we left the fishing gear at home and all canoed together, including one memorable trip to southern Florida where horseshoe crabs stubbornly tried to mate with our shiny aluminum craft.

I tried fishing with the other family member, our dog Hyper. Big mistake. Hyper was a medium-sized mutt with a big heart and the endearing habit of growling at prefabricated tube-biscuits as he buried them in the front yard. I took him to Bluff Lake, a beautiful, shallow expanse on the Noxubee Wildlife Refuge not far from Starkville. A canoe is not the most ideal of fishing craft at the best of times. It was Hyper's first time in a boat. Things were going well until I hooked a sizable bass a long, long way out. It hit my spinnerbait, hard. The strikes elicited by bladed lures are like electric shocks. This one was like a thunderbolt. Excited by my equally enthusiastic hook set, Hyper jumped out of the canoe to see if he could help. We were far from shore, so I had no choice but to go in after him, even as I held the rod high and back to keep the line taut. Fortunately, Bluff Lake is not especially deep. I got my shoe tips into the mud. Hyper crawled up to a scouting position atop my head. Somehow, I managed to get a soaking wet thirty-pound dog, a four-pound bass, and my own bedraggled self back into the canoe, bringing the expedition to an ignominious end. Well, not totally ignominious. I did not get skunked.

The expeditions waned as the years went by. For dedicated professors—and most are—there never are enough hours in the day. After getting a PhD, I was to find that out for myself, joining Janet in the Anthropology program at Mississippi State University in the fall of 1999. We worked very, very hard, helping to found and develop a new master's program that turned out to be gratifyingly successful, overseeing grueling summer field schools, meeting the demands of contract archaeology, doing and publishing research, doing public outreach, participating in professional societies, suffering through endless committee assignments, developing classes, teaching, and grading, grading, grading. Especially after the kids had left home, work consumed far too much of our lives. Eventually, the walls closed in such that something incredible happened. I *stopped fishing*. For nigh on *seven years*. My fishing partners were aghast when I gave my gear away. It was awkward, a bit. But I had other things to occupy my mind in any case, as Janet wound up battling a number of ailments, the most serious of which was a recurrent cancer that slowly, slowly chipped away at the monumental rock of her vitality. She never yielded. It took her from us regardless.

Such a mighty heart, gone. And with it, so much of ourselves. I have a picture on the wall in the room where I write. It is from 1986. I am behind Janet on one side, David on the other, and little Nikki is on her lap. Janet is central to us all. And so she was. Dedicated, courageous, stout of will; vulnerable, shy, and easily hurt. Bookish fan of the Clash. The strength of her convictions was the strong force holding together her contradictions. Hurt of others was the thing that bothered her most in this world: injustice, her *bête noir*. Good people are rare enough. Janet was a diamond, for those with eyes to see. Lesser spirits could be intimidated by her, which just shows that they did not know her well. She was good to everyone who allowed it. Sometimes that trait bit her, as she could be too trusting. Only the cynical would construe that characteristic as a fault.

Work was no refuge from loss. Janet and I had shared an office, an Odd Couple arrangement in which she was untidy Oscar while I was fussy Felix. Entering that office after her passing became a thing of dread. We also had shared friends and colleagues whose commiseration helped, even when I did not want to hear it. Unable to face carrying on at work without her, I retired. The kids, having not been kids for some considerable time, went back to their respective lives. Alone for the first time in my truly adult life, I found myself staring into a peculiar void. When you build a life with someone and find that foundation gone, so too fall the walls of identity. *WHO I AM* becomes *WHO AM I*; question, rather than affirmation. I was lost. It would be some years before a brighter light than I could have imagined beckoned me forward again. In those years I had few sources of real solace. Bored, unsettled, and utterly worn out, I decided to see if fishing still had its old, healing charm.

And so it did. I felt it even as I was in the box store looking at rods and reels. I purchased a nice little Mitchell spinning reel and paired it with a five-and-a-half foot, medium-action Shakespeare Ugly Stik rod. Then a small tackle box and some basic lures. H&Hs! Plastic worms! Rapala minnows, the "broke-back" kind! Beetle Spins! It felt like Christmas. I convinced myself that I had been a good boy and filled up my stocking accordingly.

Keith had been one of Janet's best friends for years, and he and I had been amigos since the summer day in 1984 when he hook-shotted a full water canteen over his head, across several intervening beanfield rows, and onto my unprotected noggin. Yeah, it left a mark. During our student years, we fished his little pond outside Mathiston, a spot so productive that we took to calling it the "Barrel." A good fishing hole and a good fishing partner, who enthusiastically agreed to my request that we plumb the Barrel again. I got very little sleep the night beforehand.

The pond was as lovely as I remembered, only about a quarter-acre in size but with beautiful water and a mix of lush vegetation around the edges, from stately pines to quivering rushes to lure-greedy button-bush. Keith was fishing with his daughter's Barbie pole, which meant that whatever else happened, I could *not* let him outfish me. I stood on the bank and took a breath, eyes automatically reading signs of depth, structure, cover. When I released the bail, the feel of the double-bladed H&H pulling the line taught across my fingertip was like shaking the hand of an old friend. I cast out over the water. I had forgotten how beautiful was the arc of the line. I started reeling in, feeling the familiar resistance of the spinning blades against the water the way it feels when you ride a bike again after a long time. *Bam!* A strike on my first cast! The hook set was perfect, long-dormant muscle memory brusquely awakened. In came a small bass, only about half a pound, but fighting all the way. I gently lipped it and removed the hook, marveling at the plucky little creature before releasing it. It had been a long, long time. I cast again, and *bam!* Another strike! In my first six casts I caught five bass and had one drop off the hook just as I lifted it into the air. By the end of the afternoon, I had caught exactly fifty fish.

The Great Hiatus was over.

Obsession returned. Obsession morphed into downright mania. I fished a *lot*. Twice in the course of two weeks I woke myself up setting the hook on dream bass. I studied knots, learning that enthusiasts of this or that particular form maintain claims of superiority with near-religious fervor. I eventually settled on the Palomar, for most purposes. If the hook eye is too small, I use a uni knot; for crappie jigs, something resembling a loop knot. Those are about all I can manage with hands that have seen better days, but they get the job done. I bought a special set of pliers for removing hooks. It differs from regular needlenose pliers by being prettier and costing a lot more. I learned that the size of one's catch can be gauged by how high up on the inner thumb the skin is roughened by lipping the bass. I developed my own secret weapon: the Claude Rig. This is a little popping bug, some inches above which is affixed a small shot weight, just heavy enough to provide some casting distance and to make the fly slowly sink. Twitch that thing back, and strikes can be elicited from pretty much every species out there. I have caught three-pound bass that way, which is quite the challenge, especially on an ultralight. I decided that it was high time I learned to fillet my catch, so I bought a nice knife with a beautifully thin, curved blade and watched lots of lurid videos on the subject. I learned as a result that there is no one way to fillet fish; rather, there are styles, often handed down along family lines. I settled on a style that I found comfortable, eventually getting my time down to three minutes per fish. That is not bad, unless one routinely brings home fifty or more fish. Having learned the basics the hard way, I invested in an electric filleting knife with which I can clean a fish in under a minute. But I still have the analog version, should civilization plunge into the abyss and no more electricity be forthcoming.

I perused fishing catalogs for stuff I did not need. I even researched computer fishing games. Alas, they turned out to be just as boring as they sound. Too bad. I remember one cool handheld game I got for Christmas back when the kids still lived at home. It had a little rectangular screen atop a handle that felt nothing at all like the real thing. One held this device and made a casting motion; a readout on the screen indicated how far the virtual lure had gone. The handle one turned to reel in actually felt pretty familiar, and there was a little wheel on the unit you rotated with a thumb to adjust the drag. Even better, when there was a strike, you actually felt it! And you could feel the fish fighting as you brought them in! This was accomplished via some sort of haptic technology. "Haptic" is derived from the Greek word *haptestai*, meaning "to touch." Whoever designed that game certainly was touched, for its awesome features were radically compromised by the ridiculous programming. In the same body of virtual water, one could catch virtual bass, snook, char, pike, bream, and swordfish. Well, actually you could not catch a virtual swordfish; you could only lose them. It was simply impossible to reel them in fast enough while simultaneously manipulating the virtual drag to keep from breaking the virtual line. Impossible, that is, until David and I figured out a hack, which was to remove the reeling handle and attach a variable-speed power drill to the exposed shaft. As soon as a swordfish was hooked (the game graciously informed you about the species as soon as they hit, another bit of programming idiocy), David high-torqued the drill while I frantically thumbed the drag up and down to keep the stress indicator within acceptable limits, shouting, "Slower! Faster!" as the situation warranted. Our concentration was that of paramedics working a heart failure case. It was intense. After about a ten-minute battle, we finally landed that sucker. Mission accomplished. If anyone thinks they can boat an actual swordfish with this technique, I will supply the drill.

Far more than ever before, I began seriously to study fishing. I had a lot to learn. First, I had to conquer an aversion to using the internet for research, a reluctance instilled by having an undergraduate student once turn in a paper asserting that the Aztecs had sewn live chihuahuas into the chests of their war captives. I mean, really. Fortunately, YouTube is replete with really useful videos. I learned a heck of a lot by watching "Bama Bass,"[1] for instance, a series put out by Stephen Russell, an amazing fisherman blessed with one of those rare voices that one can listen to all day. He is a master of all lures, but especially of frog fishing, which he makes look easy. His comely wife Liz often outfishes him, which may be the very definition of a successful relationship. "Creek Fishing Adventures" by east Tennessee stalwart John Dalton is another favorite. That guy will drop a line anywhere, and his genuine appreciation of the endeavor is endearing.[2] NDYakAngler[3] puts you right in the seat of a fishing kayak with Matt Nelson, who draws feisty smallmouth bass like metal shavings to a magnet. "Intuitive Angling" features Randy Blaukat, a professional bass fisherman and amiable iconoclast who shares insider secrets of the trade in Penn and Teller fashion.[4] That and his propensity to criticize have flipped a lot of people's switches, but no one could say that his heart is not in it. Hannah Barron, the scantily clad, impressively buxom Noodling Queen, has sent many a redneck into blissful cardiac arrest with her handgrabbing videos,[5] emerging dripping wet from the muddy water with her arm disturbingly deep down the maw of a catfish that weighs half as much as she does. That is a psychological study waiting to happen. TikTok also has lots of interesting videos posted by some awfully good anglers. Two of my favorite series are Hood Fishing with Rasheen Bailey[6] and Rain Outdoors with Rain Williams.[7] Kay Fishing with

Kara Moss[8] is worth watching just to see her beautiful smile. You do not have to be an angler to appreciate any of these characters. They are treasures, mildly to wildly eccentric, hugely entertaining, and outstanding as teachers. What emanates from their videos, besides true skill, is the pure joy of fishing. Those folks cannot *wait* to get out there and wet a line. Which is infectious, to say the least.

My favorite YouTube fishing guru is Richard Gene, "the Fishing Machine,"[9] a gentleman who could be a hand model for *cutis rhomboidalis nuchae* and who has a habit of slipping into a manifestly deranged alter ego named "Elmo" about whom the less said, the better. Little ones, he calls "poot sniffers," while big ones get his "adrilogen" going. He is a hoot. But man, oh man, can that guy fish. More and more, fishing is becoming a thing of scopes, technology so advanced that not only can you see the fish, you can follow your lure down so as to target them with astonishing precision. Where is the fun in that? It is like watching someone put the presents in your stocking. Mr. Gene has a depth finder, but primarily he relies on experience, vast experience. His knowledge of patterns for a wide variety of species is simply incredible; he knows exactly what to throw, when, and where. He is a true master, conversant with all the variables that come into play; as he put it in one memorable video, "there's lots of minutiaes in it." Plus, he respects the fish, even to the point of spilling water onto the metal bench of his boat to cool it down before flipping a fish onto the surface. And when he keeps a mess, Mama Sue cooks them up in traditional Southern style, so that you simultaneously salivate and feel your arteries harden while watching. I do not think that Elmo is welcome at the table, though. I love those guys.

One effect of the social media experience was that I bought way, way, *way* too much gear. When the Bama Bass guy landed an eight-pounder on a lure that looked like a cross between a ceiling fan and a sea urchin in a tie-dyed cheerleader's skirt, I immediately ordered five. If someone got a heart-stopping topwater explosion on a gizmo that looked like a UFO and the Statue of Liberty had mated and birthed a crossdressing parade float, why, send me a dozen! It was via YouTube that I discovered the Whopper Plopper, a devilishly effective topwater lure designed by "legendary angler and TV personality, Larry Dahlberg" that "delivers a sputtering topwater disruption that is begging to be crushed."[10] Truly, there is nothing quite like a sputtering topwater disruption begging to be crushed. Whopper Ploppers come in a range of sizes and colors. I tried them all, short of musky-size; one of my favorites is the Whopper Plopper 75 in Munky Butt. Dignity becomes a second-place concern when one is obsessed. I still have a large cardboard box full of tackle from that prolonged binge.

I also bought multiple rod-and-reel combos, from ultralight bream rigs to light crappie rigs to a good, stout bass rig to a heavy duty catfish setup that might have landed a small barge. My rod holder looked like the cue stands you see in pool halls. There were no empty slots. I bought a portable electronic scale. I bought a clicker to record the tallies of my catch. And there were other tricks to the trade, other products that I learned about as my tablet and I bonded throughout the long, lonely nights. I purchased a bottle of shockingly bright, garlic-scented fish attractor, a liquid into which you dip the tails of plastic worms or other creature baits, turning them an alluring chartreuse and supposedly stimulating bites. I managed to spill the bottle in my tackle box, so that for a very long summer everything in there smelled like French cooking. I bought a fishing vest to carry extra tackle, a broad-brimmed hat and a face mask against the sun, gloves with tips cut out of thumbs and index fingers to protect the hands while enabling knot tying. Fully equipped, I looked like a street busker in Atlantis.

Sometimes this addiction paid enormous dividends. I became aware of two lures that work so well that I seldom use anything else now, despite the gazillion dollars' worth of junk cluttering up my tackle boxes. One is the Ned Rig, a mindbogglingly simple setup created by Ned Kehde, an outdoors writer who looks suspiciously like Dr. Sivana from the old Captain Marvel comics. Like Sivana, he is a genius, but one who has turned his powers to good. The Ned Rig is basically a stick worm four or five inches long, threaded onto a robust hook with a mushroom-style jig head. A stick worm is just a straight, stubby thing, with no flappy tail or other appendages. It hangs from the hook in the most blasé, seemingly unappealing way imaginable, looking nothing at all like any sort of natural prey. Yum Dinger is a popular brand of stick worm to use. I need a job naming tackle. But holy smokes, does that thing catch bass! There are various ways to retrieve a Ned Rig, but my favorite is to cast it out and let it sink to the bottom. Then you raise your rod tip and stand there like a foreign tourist trying to decipher the New York subway map while the lure sinks in a slow arc back down to the bottom, the worm sticking up vertically while the jig head does its humble obeisance to gravity. Drop the rod tip, reel up the slack, and repeat. And hold on, because a bass is *gonna* smack that thing. Or even better, a bass will pick it up and swim off with it, so that your first clue that a fish is down there is when your line starts moving sideways on its own accord. That is a moment to live for. I have caught very many, and some really big, bass this way, in all kinds of water, in all seasons, in all kinds of weather conditions. Thank you, Dr. Ned!

The other magic lure does not, at least in my experience, catch as many big bass, but in terms of sheer numbers of fish, it simply cannot be beat. This is the Wacky Rig. The origins of this aptly named tidbit are obscure, to say the least; one online wag suggested that it has been invented by every three-year-old who ever has been fishing. What you have is a hook, the point of which you stick through the middle of a plastic stick worm. That is all there is to it. Weightless Wacky Rigs work very well, but I like to put a small, bullet-shaped worm helmet atop the hook to increase casting distance. This floppy little combination, which costs almost nothing, is devilishly easy to fish: cast, let it sink, then slowly reel it back, giving the occasional twitch and a pause or two. There is no quicker way to fill up a stringer. Legendary angler Gary Yamamoto developed a worm that works particularly well, called the Senko, which sounds like a cool name for a goldfish, or maybe a Chihuahua. Almost as cool as Munky Butt.

Both for safety's sake and because I enjoyed the competition, I usually fished with other people, although that took a lot of cajoling on my part as everyone seemed to have a job and/or something more in the way of a life than I. When I could not coerce anyone into coming along, I would go anyway, usually under the pretense of scouting out some new fishing hole but really just because I was hooked. I began to realize that I had a problem one day when I was doing some clean-up work at the university. It was *hammering* down rain, the hardest and most prolonged deluge I had seen since Camille pushed her freakish way up the state back in sixty-nine. As I sat there, putting ancient potsherds into clean, labeled little plastic bags, my eyes kept wandering to the window.

"Now, what would happen," I wondered to myself, "if one were to run a buzz bait across the water in this kind of downpour?"

Dammit. I drove home, my clothes soaked through just from getting in and out of my car. My dogs looked at me with something less like pity and more like alarm as I grabbed my gear and headed back out the door. I drove, sodden, about thirty miles to park alongside a lonely gravel road cutting through the piney uplands in the hill country northeast of Mathiston. There were two ponds on the property north of the road that I had permission to fish. Both were extremely small, but both had yielded nice bass in the past. The rain had slacked up not one whit. I got out of the car, got my gear from the hatch, and headed into the dense woods. I got lost once, as visibility was limited and so much of the ground surface was covered in water that I could not identify the subtle landmarks via which I usually navigated. The weather grew even worse. There is wet, and there is wet. I was *wet*.

Eventually I found my way to the larger of the two ponds, a remote and sweet little pool, perhaps a quarter of an acre in size. The water looked like it was on rapid boil. So much rain was pouring from the brim of my hat that it was like looking through a waterfall. Undeterred, I flung a large, black buzz bait almost clear across the pond. As best as I could hear against the downpour, the lure made a satisfying commotion as I reeled it back in, but no strike was forthcoming. I cast again. And again. Nothing. It became hard to concentrate as rain lashed the surrounding vegetation, ten thousand brushes beating the tar out of five thousand snare drums. Distracted, I screwed up a back cast and hung my lure high up in the limb of an oak tree. There was too little room amidst the surrounding vegetation to finesse its release. After a couple of desultory twitches of the rod tip, I grew angry; angry at the weather, angry at the lack of bites, angry at myself for being such a dunderhead. I snatched, hard. The rod broke. Just then, lightning struck so close that I literally felt the shock wave. I suddenly remembered that I had left my cell phone in the car because of the rain. I suddenly remembered that no living soul on the planet had any idea of where I was. I suddenly felt a very strong urge to call it a day. For a change, I listened to common sense and splashed my way back to the car. Experiment, complete; findings, negative.

That ridiculous expedition was not the only dangerous thing I did during those obsessive years. The same man who owned the pond where I had nearly gotten smacked by lightning owned another one deep in the woods of Oktibbeha County. Again, there were no trails, nor was there any road front on the property. To get to the pond, one had to park at my friend's house and cut through the woods, down steep ravines and up narrow ridges. The last ridge had a powerline right of way allowing a clear line of sight to the back yard of a neighbor's house. The neighbor's house with a large sign out front warning visitors against dangerous dogs. My friend likewise had warned me about those dogs; they meant business and had bitten more than one person to prove it. One was a German shepherd, the other some sort of pit bull mix. They were seriously large. I never have understood why people keep dogs that are a real danger to others, but I will say this for the lady who owned them; she believed in giving them plenty of outdoor exercise. At least 70 percent of the time I made my way to that hidden little pond I had to cross the open avenue of destruction while she and the Cerberus twins cavorted in the back yard. I have no doubt that, had those brutes spotted me down the cleared alley of the powerline, they would have raced each other in a friendly competition to see which of them would first eat my face. I never turned back, but would literally hug the ground, dragging my tackle box and doing my best to keep the rod tip down without hanging a lure in the vegetation as I crossed the danger zone. Then there was the repeat journey, always with a loaded stringer in tow. It was quite the honey hole. I somehow managed to avoid detection every time, but it always was a nerve-wracking experience.

The biggest hurdle that I had to face getting back into the sport was finding good places to fish. This is a common problem because public lakes tend to stay pretty well overfished, while the larger reservoirs require some truly serious gear and years of experience to properly figure out. One advantage I had is that I will fish anywhere, including—and especially—in ponds of any size or description. There are few things I love better than casting out into a pond where I have never fished before, and I usually have inordinate good luck on those underappreciated little bodies of water. Many little ponds hold bass of decent size, and virtually all of them have a "pond monster," one finned colossus that occupies the topmost rung on the food chain. Many is the time I have dropped a Ned Rig into some innocuous little farm pond only to tie into a toad. I hooked one such monster while fishing a pond outside of West Point, Mississippi with my friend Bobby. We had been catching the usual complement of half- to one-pounders, having a fine old time, when I worked my way to the other side of the pond to drop my lure down near the roots of a solitary cypress tree. It had not hit bottom before the line took off to the side. Thank you, Dr. Ned! I set the hook and immediately knew that I had tied into a big one. I made frantic adjustments to the drag as I worked my way to the edge of the bank, fighting to keep the beast from wrapping my line around the many sunken branches in the vicinity. Alas, my last step onto what looked like solid ground turned out instead to be onto hay floating on the water. I went in up to my knees but still managed to keep the fish hooked while I struggled back onto shore. It weighed six pounds. Giants walked the Earth.

My efforts to find places to fish became shameless. I wheedled and cajoled. I offered to help out around people's farms. I started every conversation with, "Do you know any good places to fish?" I paid regular clinical visits to an acupuncturist who filled my ears with tiny needles and who eventually granted me access to her little pond. I took unwanted dulcimer lessons from a nice lady who owned a magnificent lake outside of Starkville. I bought some land in Webster County that had two small, fish-laden ponds. Brother Glenn and his wife Kathy began referring to me as a "pond whore." I wore the mantle proudly.

All such effort paid off when the Covid-19 pandemic ramped up in early 2020. It was a strange time for many reasons, one being that it was the single best fishing season I ever have experienced. I must have caught two thousand fish that year, many on solo expeditions because even my most stalwart fishing partners were reluctant to share a truck cab or boat out of fear of contagion. Besides providing welcome recreation and a source of food when I was avoiding the grocery store as much as possible, fishing during the pandemic brought other benefits. For a very long while, toilet paper was very hard to come by. Fortunately, I was good friends with a couple who lived in Starkville and who had a surplus. Paul and Nancy liked to eat fish at least once a week for health purposes. They also preferred to avoid the grocery store, for the same reason. Two freezer bags of fillets became the bartering equivalent of two rolls of TP. Life finds a way.

My obsession did not stop at fishing rigs and tackle. Having hit my late fifties, I decided it was high time that I had a boat. Bank fishing is fun, but it is a lot of work to cover any real amount of water, and carrying all your gear gets old pretty quickly. Boating is another world altogether, especially with the right craft. I did not want anything big, at least not at first: small boats are more intimate by far. To be afloat in a small boat is to be cupped in the hands of a vastly larger spirit, briefly visible in the early morning mist rising from the water. No true angler can be unmoved by such a sight. The spirit-craft I was looking for was a two-seater pontoon boat, small enough to put in the back of a pickup truck and take anywhere without bothering with a trailer. I scoured the internet for options and my own spirit soared when I found what I was looking for. It was a Twin Troller X10, a captivating little ten-foot-long craft with two electric motors hidden within slots running underneath the length of the hull. Not a pontoon boat, really, as there was hull underneath your feet, but the same basic idea. The motors, which were operated independently by a dual-pedal foot control, were reversible, so that you could keep your hands free and go forward, back up, or turn on a dime. Although small, that boat was virtually unsinkable, as it would "suck down" once water was pulled into the recessed slots. The innumerable YouTube ads I watched proclaimed it to be, "The best mini bass boat ever!" Reality, alas, was to wear a different face.

The boat was delivered unassembled in a series of huge cardboard boxes lashed to wooden pallets. The eighteen-wheeler could not get up my little street, so the truck driver obligingly dropped the pallets off in the yard of a forbearing neighbor on the corner. Eventually I got everything to my own yard, one piece at a time. One heavy piece at a time. Assembled, she was a beautiful thing. I called her the *Cute Little Bugger*. Moving her around in my driveway was not too difficult, as I had sprung for an oddly shaped dolly specifically designed to slide up under the boat. That worked okay, as long as the ground was pretty level and not too soft; otherwise, the whole thing just tipped over on its side. Hauling her up a slope was exhausting in the extreme. To load her into the back of my pickup truck was another trial. I would heave the front up and onto the tailgate, then lift the back and slide the whole thing forward in the bed in a single, smooth, herniating motion, an activity I was not sure I would be able to undertake during my rapidly approaching golden years. Once in the truck, though, a couple of bungee cords was all it took to secure the *Cute Little Bugger* in place. Performing the operation in reverse was tricker, as it was hard to slide her off the truck without initiating a catastrophe. Accordingly, I embarked upon a series of efforts to engineer a solution, the main result of which was a succession of Darwinian reminders that I am in no way an engineer.

My first stratagem seemed sensible enough, which was to obtain a couple of heavy-duty four-wheeler ramps, the idea being that I could set them on the tailgate equidistant with the dolly wheels and just roll that ol' boat on and off the truck like nobody's business. That actually worked pretty well when it came to loading the boat, especially if I had a buddy to push or pull on one end, but unloading was a very dangerous affair, as upon its descent the heavy boat on its wheeled dolly would simply pull one at high speed into the void between the ramps. I tried having a fishing partner brace the stern, but that just threatened to decrease my supply of fishing partners. The foolishness of the operation became apparent early one morning when Bobby and I were unloading in a parking lot beside the Tombigbee River outside of Columbus. We had fished the river in the *Cute Little Bugger* once before, which was not the wisest of moves in any case. Things were okay as long as we hugged the shoreline, fishing little creek mouths. But an enchanting island in the middle of the river, and enticing sloughs on the far side, tempted us out into the deep water, where I got leg cramps trying to push the motor pedals through the floor while that little boat inched its way across a very strong current. The Tenn-Tom Waterway was constructed to allow barge traffic between the Tennessee and Tombigbee Rivers. Had a barge come through while we were nosing our way across the river in that little boat, we would have been sitting ducks. Accordingly, our plan for the day was to hug the bank and fish little creek mouths. But first, we had to get the boat unloaded. Gingerly, I rolled her forward on her dolly and onto the ramps, to the point where gravity, that harsh mistress, began to take over the operation.

"Here she comes!" I yelled to Bobby, suddenly remembering that neither of us was a spring chicken anymore. I heard him grunting with effort, even as the boat began rolling still faster down the ramps. A bit of lightning calculus crackled through my brain, the sum of which was that my friend was dearer to me than my boat, cute though she might be. Desperately, I mounted the ramps while heaving against the dead weight. The ramps jostled, bounced . . . and came down, along with me and the boat, all landing with a loud bang on the hard asphalt. I lay there for a while, trying to find my voice within the thicket of pain, while Bobby looked to see if any bones were protruding through my clothing. I was very lucky to come away with only some major scrapes and bruises. We fished anyway, of course. We grow 'em tough in the Hills, if not particularly smart.

Loading and unloading would have been fine on a trailer, but that was not what I was after at the time. I put my mind to the problem, which usually is a mistake. I attached a come-along to the front of the truck bed so that I could cable the boat up or down the ramps with ease. Alas, much like Leonardo's flying machines, the operation worked better in theory than in practice. The slightest deviation in direction, hard to avoid with an unwieldy boat on a small-wheeled dolly, caused the whole affair to slide over the side of a ramp, initiating a crisis. Plus, after about three or four episodes, the front panel of the truck bed began to bend outwards. That little boat was *heavy*. The come-along maneuver not really working, I began to devise yet another solution, which was to install some sort of boat lift in the rafters of my garage so that I could back the truck in and raise or load the boat as occasion demanded. Fortunately, things never went that far, as other problems convinced me that it was not, at least for me, the "best mini bass boat ever." For one thing, there just was not any place onboard to stow tackle boxes or other gear. For another, while the boat was indeed virtually unsinkable, if both passengers were fishing on one side she leaned in that direction, so that it always felt as if one were going to slide out of the seat and into the water. The final solution I engineered turned out to be the best one, which was to sell the *Cute Little Bugger* to a guy who lived on a private lake where she could just stay tied up to the dock. Fortunately, I had not had her name painted on the hull. For him, I hope she was indeed the perfect little craft. For me, it is a wonder that no serious injuries occurred.

As if I were not obsessed enough already, my friend Jeffrey and his brother Ron took me out to the mouth of a tributary stream feeding the Tombigbee River and introduced me to real crappie fishing. It was late February, still early enough to be quite cold, when we launched at a terrifying little private ramp and made our way upriver to the stream mouth just at dawn. The setting was beautiful, blue herons framed against the early morning light by moss-draped cypress limbs while spirit-breath rose off the water, like a Walt Kelly Okefenokee drawing. Because the river was up, the stream mouth was full, a broad expanse of water pierced by stumps like the broken masts of sunken vessels. Many more were barely visible as dark spots beneath the surface. That made for some interesting moments, as more than once the boat ran up over a submerged stump and came to a lurching halt, requiring the three of us to sway in conga-line unison in order to dislodge the craft. But those same stumps provided excellent cover for crappie. Ron graciously attached a leader and a slip cork to my line, at the bottom of which I affixed his recommended jig, the "Electric Chicken." After sticking a nearly phosphorescent crappie nibble onto the barb of the hook, I cast out, following Ron's direction to slowly, slowly reel the line back in while the current and little wind-driven waves did the work of playing the jig. The technique worked to perfection. It was a beautiful thing to see the cork slide under the water, and the crappie fought like demons. With beginner's luck, I caught more than my fair share, but was happy when, on our way out, Ron boated the catch of the day, a toad that might have gone three pounds.

Pleased beyond measure with that expedition, I had little trouble convincing my friend Allen to bring his boat and try the same creek mouth. With exactly one previous trip under my belt, an expert guide I was not. Nor was I an expert at anything having to do with boats, as the *Cute Little Bugger* debacle should have warned me. We hit the same little private ramp at dawn. The river was way, way up, covering most of the ramp, which descended at a steep angle straight into the river with nothing to break the flow of the very strong current. A thick layer of oozy mud covered that part of the ramp still out of the water. I knew enough to know that wet mud on concrete is very slick, so I did not want to stand there holding the bow rope while Allen backed the trailer into the water. The large concrete post adjacent to the ramp on the downriver side offered a solution. I perched upon it like a late-middle-aged seagull, holding what suddenly seemed to be a very short length of rope. Allen backed up quickly, hitting the brakes before he got too far down the ramp. The boat shot off the trailer and into the water. It was indeed a very short length of rope. I was yanked right off the post, landing on the ramp and sustaining considerable damage as a result. As soon as I stood up on the slick mud surface, my feet went right out from under me, so that again I hit the concrete, hard. That happened twice. By the time Allen made his way back to the ramp, I was a soaking, mucky, bloody mess. But at least I had not let go of the rope. Concerned and somewhat chagrined, he asked me if I wanted to carry on. By all means, I replied.

Not far downstream from the concrete post was a crappy little wooden dock in an advanced state of disintegration. I knelt there and held the boat while Allen stepped in, learning a valuable lesson in the process: it is best not to have your thumb between gunwale and dock whilst someone is boarding. The pain when my digit was mashed between the two hard surfaces was excruciating. The skin ruptured; blood poured. I have no doubt that a bone was cracked. It was a revolting sight to behold. Concerned and somewhat chagrined, Allen asked me if I wanted to carry on. By all means, I replied. We grow 'em tough in the Hills, although life expectancies are correspondingly short. Fate provided some compensation, as I wound up catching the biggest one of the day. Totally worth it.

Looking back at that obsessive period I find little to regret, outside of the credit card bills. Did fishing fill the abyss of loss? Of course not. Did it help? Yes, it did. Fishing provided joy at a time that might otherwise have been largely joyless. It gave me something to focus on during long hours in a house that echoed with melancholy memories and rewarded me with a host of newer, happier ones. I *caught* those dream bass. And many, many real ones as well. As obsessions go, it could have been worse. I might have taken up pop line dancing.

DOUBLING UP

Every redneck's dream is to write a song
and have it go on a fishing show.
—JOHNNY VAN ZANT

My brother Glenn is the stuff of which legends are made. He pushes his way through life the way an icebreaker makes its own sea lanes. He has had eleven—count them, *eleven*—hip surgeries, and as of this writing a twelfth may be in the offing. One such episode left him lying in bed for days without an upper femur while they sculpted a replacement. "That hurt," he said. For Glenn to admit that something hurt meant that most mortals would have made out their wills and put the morphine pump on speed dial. Ordinary scrapes, cuts, and contusions Glenn brushes off like insect bites. More serious misfortunes, like snakebite and falling off a bridge, slowed him down but did not stop him. Multiple operations on his hands were, he grudgingly admits, bothersome for the very good reason that they make it harder for him to operate a reel. For among his many other fine qualities, Glenn loves to fish. That is one of the many reasons why I love him.

Glenn's devotion to fishing began early and was at least partly attributable to yet another affliction, namely that he could not go swimming without developing serious ear infections, every single time. Thus, over the years while the rest of us boys splashed about in creeks, cow ponds, and other water bodies with dauntingly robust bacterial loads, Glenn would cut a solitary figure on the far side, rod cutting the air with studious regularity, hoping, perhaps, that we inadvertently were herding fish his way rather than spoiling the endeavor altogether. That sort of obstinate optimism in the face of adversity is another of his fine qualities. Although sometimes I think that it is locked in battle with his survival instincts, the winner of which has yet to be determined. I am hoping for at least a draw.

That image of Glenn, standing alone at a distance while we youngsters cavorted in the water, has remained fixed in my mind for many decades. It exemplifies his unique position in the family lineup, the Middle One sandwiched between the Older Boys and the Little Kids. His acting out, another perennial trait, perhaps was an effort to fit in with one or the other group. I think we all wanted to claim him. One summer night, when I was about nine, we Little Kids and our cousins Butch and Mike "camped out" in a wooden playhouse Uncle Billy had built on their property outside of McCool. Whether because he was being punished for some transgression or simply because there were not enough beds for everybody in the big house, Glenn spent the night with us kids in what was basically a small, plywood shed. He was in excellent form that evening, as he turned off the light and entertained us with a graphic, blow-by-blow description of war between Earth and "the Clod Planet." I leave it to the reader to decipher the euphemism; a major clue is that we young'uns were laughing to the point of asphyxiation as Glenn illustrated his tale with the flashlight beam on the wall. Clods whizzed back and forth across the cold, pitiless void of space. Repeated attempts at negotiation were futile. Earth took a beating. Those Clod Planet people were full of it. Ammunition, accordingly, was inexhaustible.

I did not fish with Glenn during my early rod and reel days, as contrary circumstances prevailed. Outside of school and chores, I was a more-or-less carefree teenager, often on a pond or creek bank while he was away working various manual labor jobs and falling off bridges. Years passed, as years do, during which I managed to fish with him only very occasionally. It was during my time as a widower that we reconnected, as Glenn was always there when I needed him. And thanks to that reconnection, I learned to love fishing even more, in part because fishing with him doubles as a way to hone my own survival instincts. Fishing with Glenn is, to put it mildly, always an adventure. One episode involved an out-of-fuel boat on the leading edge of a hurricane.[1] That was a moderately terrifying experience, but at least we did not get skunked.

Before discussing Glenn's gear, some contextualization is necessary; to wit, acknowledgment of the fact that we grew up poor. Really, really poor, such that bartering, wheeling and dealing, and strategic cajolery were early engrained as part of our cultural repertoire. Glenn is a master of all such arts. Once, in his early married years, he drove past a house where an old tire lay abandoned in the yard. A tire that looked like it just might be a fit for his rims. Even used tires cost money; this one was just *sitting* there. Excited by the prospect of some useful road rubber, with a characteristic, disarming forthrightness he knocked at the door and, when the owner answered, asked if he might have the tire. "Sure," said owner answered. "We need to get it out of the yard anyway. Mind the ants." Fire ants, as it happened, that had built a stately mound over the object of Glenn's desire. Said ants were not at all pleased when their real estate was rearranged by an uppity biped. Glenn's pains were rewarded though, as the aged rubber torus did indeed fit his rim. He got two weeks' use out of it before the weatherbeaten thing disintegrated while he was driving. Such were the spit-and-bailing-wire skills that carried us all through some very hard times.

Glenn's collection of rods and reels is rather like a used car lot; some once-good models, some not, often of the hybrid variety, all of them showing considerable wear and tear, piled up in no particular order and almost always in need of restringing. As it happens, that is a pretty standard situation for a backwoods fisherman. But his tackle box is the stuff of wonder, a big, ugly plastic thing full of all kinds of hook-bearing bric-a-brac, some shiny, some rusty, like the carrying case of a zombified Mary Kay salesperson selling swamp-to-swamp accoutrements to the Walking Dead. Most of the contents were acquired due to an unexpected windfall, when Glenn espied a large plastic barrel chock full of ridiculously discounted tackle at a K-Mart in Glendive, Montana, holding a going-out-of-business sale. "Glendive" it was, as he repeatedly plunged into the barrel, going back every day for two weeks, exchanging cash he earned from small-engine repairs for scores of packages of leftover tackle that more knowing fisherman had passed up with scorn. There were creature baits with dubious appendages vaguely suggestive of Venusian sexual organs. There were clear, plastic corks that could be filled with water to enable long-distance casting thanks to the additional aquatic payload. There were little rubber things that might have resembled crickets to a participant in an LSD trial. There were artificial frogs that were far more artificial than frog. There were rubber rats. *Rubber rats.* My personal favorite is an enormous, rubber stag beetle, a preposterous brown and yellow thing that, if nothing else, might concuss an unsuspecting fish. Glenn is inordinately proud of that particular find; it resides in his tackle box as a "secret weapon" that he can call upon should he have the need. I am sure that day will come, and I would take no bets upon the outcome. In fishing, as in the rest of life, things do not always go by the book.

Amidst this unlikely menagerie once were scores of vaguely fish-shaped, silicone swim baits of a garish, red-and-yellow two-tone color, molded around cheap hooks beneath narrowly conical, red-painted jig heads. I would have left them in the barrel. Glenn bought them by the score. One late winter morning he and I were fishing in his little pond, having no success whatsoever. Pre-spawn bass are finicky creatures, to put it mildly. On that particular morning, we had tried H&Hs, Rapala minnows, Beetle Spins, plastic worms, and any number of other standard enticements, with nothing to show for our efforts. With a sheepish grin, Glenn attached one of the little fish-shaped lures and cast it out. Almost immediately, he hooked a bass. Never one to be proud where catching fish is concerned, I cadged one of the silly looking things and gave it a try. A bass *slammed* it. And so it went, vicious strike after vicious strike, the flimsy silicone bodies of the "Wonder Lures," as we named them, disintegrating after two or three such powerful hits. The fish seemed to be genuinely angry at the things, doing their best to obliterate them. We nursed those Wonder Lures along for months before we finally ran out. Due to a wayward cast in high wind, my last one wound up high in a pine tree that I seriously considered climbing, even though my tree-climbing days are well past. The lure was that good. I have since scoured the internet and sent out multiple entreaties on social media trying to find exact replacements, to no avail. While red is a go-to color for pre-spawn bass, nothing else I have tried has come remotely close to matching those silly, cheap rubber things that Glenn secured from a moribund K-Mart in eastern Montana. If anyone out there can hook me up with a supply, I will clean your fish for you, for life.

For years, we fished from Glenn's boat, a vintage fourteen-foot aluminum craft that had seen far better days when he acquired it. For a while, this fascinating artifact was propelled by a thirdhand trolling motor that pushed us forward with reluctance while complaining angrily in a remarkably loud staccato bark. Sometimes the motor quit, just to assert its independence, I suppose. I thought of several names for that legendary craft: the USS *Submersible*; *Dauntable*; *Collander*; *Lead Balloon*; *Sinkready*; *Swimnought*; *Poseidon's Revenge*; *Cinderblock*; *Bottom Feeder*; *Footbath*; *Aquabrick*; *Sponge*. In typically elegant fashion, Glenn just called her *Old Leaky*. It was an apt moniker. We routinely plugged the more obvious, geyser-like holes by jamming sticks into them, but leaks along the aluminum seams were harder to deal with. Prior to an expedition, Glenn would put *Old Leaky* up on sawhorses and use a hose to fill her up, marking the more startling freshets to be covered later with liberal spritzes of Flex Seal. The bottom of the boat eventually had more layers of asphalt than a blacktop road outside the county supervisor's house. Centuries from now, archaeologists will be astonished when they uncover a full-scale boat replica made entirely of spray-on sealant. Yet water still found a way. Fortunately for me, I rode in the front of the boat, while most points of entry for the water bodies we were supposed to be floating upon were in the stern, which Glenn euphemistically referred to as his "live well." When water topped his ankles, he bailed it out with a Tupperware bowl carried for the purpose. On windy days, we spun about on the water like a leaf in a stream while Glenn simultaneously bailed and piloted. I fished while he bailed, which he uncharitably called "cheating." Well, yeah.

Old Leaky was borne upon a trailer that was, if anything, even more decrepit than its dubious burden. Presumably the trailer lights once worked. I am certain that, on one occasion, the "safety chain" was a short length of medium-gauge wire closed at the ends like a twisty-tie. Fortunately, we never went far. For years, Glenn pulled that trailer with a battered white pickup truck sporting a pair of bull's horns mounted prominently on the hood. It truly was a sight to see.

There is a certain madness that seizes the brains of anglers when the bite is on, especially when fish are hitting the top, feeding on insects floating on the surface of the water and/or slamming schools of fleeing shad. When bass are really hitting the top, it looks like someone has dumped a load of gravel into the water from a great height, a phenomenon guaranteed to overstimulate the amygdala. The resulting derangement can lead to some questionable decisions of the life-threatening variety. Twice, Glenn and I launched *Old Leaky* into waters where they were hitting the top, quickly discovering that in our eagerness we had neglected to insert the drain plug. Once we went to bank fish at Paw Jim's Pond, a modest little water body way out in the country south of Stewart. It is the very same pond where Glenn fished alone while we younger kids swam and cavorted so many years ago. Back then, the pond was well-maintained around the edges, so that one could fish the perimeter uninhibited. Nowadays, the shoreline is overgrown with thick, viny growth, severely limiting access. Glenn and I stood shoulder-to-shoulder in a small gap in the tree line that morning, discovering immediately that the bass were biting like nobody's business. Frustration mounted as the prime water remained well out of our limited casting range. To make matters worse, they were hitting the top. Our amygdalae took note.

Exasperated, we hastened back to Glenn's house to get the boat. Returning to the pond, he backed the rig into the water but, oddly, *Old Leaky* refused to stir from her resting place on the trailer. Belatedly, Glenn checked the drain plug. Rather surprisingly, it was in place. Perhaps *Old Leaky* required some assistance? We waded in, each to a side, took hold of the gunwales, and pulled. *Old Leaky* refused to budge. A salient point worth repeating here: they were *biting*. We pulled harder. She moved! We pulled harder still, ignoring the increasing resistance until Glenn suddenly noticed the two short, stout, rubber bungee cords holding the bow of the boat to the trailer. Those were now stretched tight, exactly like a giant slingshot. Had we relaxed our grip at that point, *Old Leaky* would have been launched, all right, not into the pond, but like a scruffy aluminum ICBM right through the pickup's rear window. Holy crankbait! We eased her back, released the cords, and had a fine morning. I caught a three-pounder on a Whopper Plopper. Life was good.

On what I thought might be the last voyage of *Old Leaky*, I brought along a technological marvel, a new lithium-ion trolling motor battery. Standard lead-acid marine batteries might last half a day in heavy winds, when the boat's position had constantly to be adjusted, and as we carried a lot of additional liquid weight between bailing episodes. Plus, there was the hernia factor to consider, as those batteries are ridiculously heavy. The battery I brought, while expensive, weighed less than half what a standard battery weighs, and, if the advertisements were to be believed, would last several times as long. We both were skeptical, but it was a good test case, and I figured that, in the event the battery did not live up to its billing, I at least would have a valuable resource for bartering when the post-apocalypse lithium wars began. To our gratification, it worked better than we could have dreamed, pushing *Old Leaky* through the water with smooth vigor and not showing any signs of weakening after a windy morning's hard fishing. I decided that bartering was out of the question: come the apocalypse, I will defend that sweet, sweet trolling battery to the death.

Inevitably, the day came when *Old Leaky* had to be retired. Sort of, anyway, for Glenn hauled her down to the little pond adjacent to his house, a pond small and shallow enough that, should *Old Leaky* go to the bottom, he could simply float-walk back to shore. A couple of years back, I accompanied him on an expedition there. I thought he was making an inordinate amount of noise moving us around. That turned out to be due to the fact that he was using a flat-bladed shovel as a paddle, his storebought one having gone missing. A few weeks later Glenn called to inform me that this makeshift device was now at the bottom of the pond, as he inadvertently had slung it a little too hard while trying to smack a moccasin that wanted to climb into the boat with him, leaving him to hand-paddle *Old Leaky* back to the bank. That must have been a sight to see. But, as he proudly informed me, at least he did not get skunked. To anchor *Old Leaky*, he used a trailer ball hitch tied to a rope. Never say never with Glenn. Eventually he attached his oldest trolling motor and just left it there on the boat, covered with a garbage bag against the elements. It worked very well to push the aged craft around the little pond, except that the first two of the five forward gears refuse to cooperate as designed, so that you had to start out in third gear and hang on against the sudden acceleration. The cables grew so hot that you had to be careful not to burn your fingers when removing them from the battery. Should a fire have started, I expect that Glenn would have begun cooking fish right there in the boat, reaching into his "live well" to get 'em fresh. That boy don't quit.

Old Leaky was replaced with a used jon boat Glenn picked up from someone over in the Delta. It is a fine boat, a solid, well-made twelve-footer that does not leak and that moves very well under the power of another used trolling motor Glenn obtained in exchange for a heavy-duty trailer hitch. One might surmise that our misadventures accordingly would have come to an end. Anyone laboring under such a supposition has not been fishing with Glenn. We hit a private lake near his house at daybreak one day in June, a fine, overcast day holding the promise of a legendary haul. That lake is the sweetest of honey holes, chock full of quarrelsome little bass that the owner wants culled, with a few big ones lurking amongst the multitudinous dinks. The three most beautiful words in the English language are "I love you." The four most beautiful words in the English language are "my pond needs fishing." Granting permission for us to fish there was like throwing meat to hungry sled dogs. We have doubled up there as many as seven times on a single trip. We almost always come away with a mighty haul, especially from the upper end where thick beds of aquatic grass provide a mecca for crawdads, minnows, and well-fed bass. Such grass will quickly choke a trolling motor, so we always bring a paddle for maneuvering in that part of the lake.

Except for this time, for the paddle had again mysteriously disappeared. I failed to find it in the bed of his pickup truck, but I did find a long, wooden tomato stake that we could use to pole ourselves out of trouble in shallow water, at least. I put it in the boat, and off we went. For that remarkable lake, the bite was rather slow that day; we caught only fifty between us, most on the ever-dependable Wacky Rig. The best bite was at the very head of the lake, right where the creek channel came in. We could see the narrow channel clearly, as it was surrounded by thick beds of grass choking the shallower water to either side. Carefully, Glenn navigated us up the channel. I managed to catch a couple of relatively robust bass out of the grass using a topwater frog. Ba-*loosh*! Glenn caught one right at the mouth of the large culvert where the creek entered the lake, a long-term goal that brought both of us great satisfaction. As our fish baskets grew weightier, we were mighty pleased with ourselves. A good time, as the saying goes, was being had by all.

Until the trolling motor conked out. Just to show that it could.

A tomato stake works pretty well as a gondolier's pole as long as the water in which one floats is fairly shallow. We were at the uttermost end of a thirty-five acre lake. There was a whole lot of water between us and the boat ramp, most of which greatly exceeded the depth limit of our impromptu propelling device. Still, there was nothing for it. We lifted the heavy fish baskets into the boat and I began paddling as best I could with the stick, which turned out to be not very well at all as it was only an inch and a half wide. Glenn fished while I stroked, which was fair enough given that he had been working the trolling motor against the wind all day. I still accused him of cheating, of course. He suggested that I get us up to trolling speed. I suggested he take a swim, bad ears or no. Fortunately, the wind was blowing more or less in the direction we wanted to go, so that we made surprisingly good time as I heaved three times on this side, three sides on that, pausing to rest whenever Glenn hooked a fish. I was pretty well spent by the time we nosed up against the shore next to the ramp. I unloaded the boat while Glenn backed the trailer down into the water, at which point the lack of a functioning motor became again an acute factor in our endearing little equation of woe.

Ordinarily, one loads a boat by lining it up just so and using an outboard or trolling motor to thrust the craft up onto the trailer, cutting the power at just the right moment to avoid ramming the thing up over the bow stop while maintaining enough forward momentum to come to a secure rest on the bunks. It is an art, especially when wind and/or current are pushing the boat to the side while you are trying to go straight forward. The wind was blowing perpendicular to the direction I wanted to go that day, as I stroked that skinny little tomato stake like a hero. I managed to get the bow of the boat onto the trailer, more or less, at which point I leapt forward and tossed Glenn the pull rope. Which broke almost immediately. Unflappable as always, he waded out to his waist and pulled the boat up partway onto the trailer. I clambered out and made my way like a tightrope walker along the narrow trailer tongue to the truck. Having not yet gotten around to attaching a hook to his winch rope, Glenn tied a knot around the handle at the bow. Glenn is good with knots. So is my stomach, which executed a double dragon loop as I noticed in passing that the locking lever on the trailer hitch coupler was secured in place with nothing more than a twisted metal pin flag. Somehow, we made it home safely. The whereabouts of the paddle remains a mystery.

At times, we have used other boats. One day we fished the Barrel, where the Great Hiatus had come to an end. I was in the bow of a little ten-foot jon boat, sculling with a small oar, moving us around the bank line. The only seats were the flat metal benches standard in such craft. After a while, I noticed that the arc of Glenn's cast had become strangely flattened and that his lure was hitting the water less with a "plunk" than a "splat." I turned to behold my older brother lying flat on his back in the middle of the boat. The lack of support having played havoc with his compromised skeletal frame, Glenn was casting across his chest from a supine position without even being able to see the water, reeling, hooking, and boating them by feel and enjoying himself immensely all the while. Obstinate optimism in the face of adversity, indeed. That boy don't quit.

It was thanks to Glenn that I was able to land my PB bass. We were at a family reunion when the partner of a cousin told us about the double-digit monsters he was catching in his little pond outside of Carrollton. We were incredulous, perhaps because, due to years of playing guitar in juke joint bands, our interlocutor described his questionable exploits in a voice that could crack an I-beam. Our stunned ears nonetheless perked up when our cousin chimed in, mentioning in all innocence that fishing the pond with a cane pole was hard because "those big fish come up from under the dock and eat the bream while I'm pulling them in." Then her partner pulled out his phone and showed us pictures. Holy jerkbait! He had not been exaggerating. They were *toads*. Toads which, he blithely informed us, had been released back into the pond. Needless to say, the invitation to come and try our luck was eagerly accepted.

Glenn and I left his house early, as the pond was a good hour's drive away. Still, it was full morning by the time we arrived, and already hot. We walked as quietly as we could out onto the tiny dock. I had been studying up on how to catch big bass, so I was prepared. I was using a Lew's Mach Crush baitcaster combo outfitted with thirty-pound-test braided line and a 5/0 offset hook beneath a quarter-ounce bullet sinker. My lure was a twelve-inch black Culprit worm with a kicking tail, rigged Texas style. I had come a long way since the days when Andy and I attached Pfleuger hooks to spliced-together badminton webbing tied to sapling poles. I took a deep breath, released it, and flipped the worm into the water right beside the outermost pier of the dock. Instantly, it was taken—*engulfed*—by something with an enormous maw. As the saying goes, it was "like a toilet flushing." The hook set was heavy, unequivocal. I staggered from the pull. The sensation was extraordinary. It felt like a booster rocket had been strapped onto an engine block. With uncustomary foresight, we had brought a net, but there was no way I could fight that monster, keeping it from wrapping around the piers, whilst kneeling and trying to net it. In strangled tones, I called for Glenn to help, forgetting in the moment that eleven hip operations rather compromise one's flexibility. Imperturbable as always, Glenn dropped to his belly and slithered across the deck, net in hand. He netted her! She seemed as wide as the dock. Once Glenn regained his feet, I prevailed upon him to take my picture, following which I weighed (eight pounds, eight ounces!) and released my toad, watching in wonder as she swam slowly back down into the water beneath the dock. Weak from the aftermath of the "adrilogen" rush, I thanked Glenn profusely for his help. He called me a "dumbass," a term of affection. He was somewhat less gracious when, two casts later, I caught a six-pounder in exactly the same spot. This one did wrap itself around a pier, somehow getting its body partway suspended in the process. Again, Glenn slithered to my aid. Again, I prevailed upon him to take my picture. And that was the last bite either of us got on that long, hot day. My poor brother, without whom I never would have managed to land those lunkers, got skunked. Sometimes there is no justice in the world.

Glenn was with me the day I splurged and bought a sure-nuff, sparkledy bass boat, a red and silver Bass Tracker Classic XL with a 50

HP Mercury motor, a live well, rod holders, fish finder, and all kinds of other bells and whistles. I should have called her *Second Childhood*. I opted for the *Diva*. I spoiled her rotten, keeping her out of the sun, lovingly cleaning her between expeditions, customizing the helm with her name in waterproof decal. In cursive, no less. The *Diva* was hauled with a long-bed Ford F-150 bought specifically for the purpose. Second childhood, indeed. She was a beautiful craft, the fulfillment of a dream I had had for most of my life. But, alas, some dreams are better left unrealized. The *Diva* turned out to be more boat than I needed for the kind of fishing I prefer. She needed bigger bodies of water than are locally common, plus I worried about her constantly. Every time I checked the rearview mirror she was still there, but one never knew. I eventually sold her, but not before Glenn and I had a major fishing adventure.

Crawling Stone Lake is a magnificent water body in Vilas County, northern Wisconsin. I had visited there and fished with Mike, a friend and former colleague who put on the dog for me over the course of a glorious weekend. He took me to favored spots where I caught my first-ever smallmouth bass, which fought like you would not believe, and rock bass, which were like hauling in scaly lumps of wet tissue paper. Mike rowed me back and forth for miles at night in a sweet little rowboat while I trolled a line, catching only another lump of wet tissue paper for his efforts but greatly enjoying the experience. I rowed in turn, the first time I ever had stroked the oars in a rowboat. It was not a smooth operation, but Mike was a patient teacher. The next day we fished a small cove sporting a nice cover of lily pads, out of which a three-pound bass exploded on the Whopper Plopper I was using. It thrashed while I was trying to release it, sinking a hook deep into my hand. I had to get Mike to yank it out with pliers, whereupon I recommended fishing as blood poured down my arm. We grow 'em tough in the Hills, if not especially astute.

A few years later, I rented a cabin on that lake and invited Glenn to join me on an expedition there. As we drove up, the *Diva* reassuringly in the rearview mirror each time I checked, I regaled him with tales of smallmouth bass, toothy northern pike, sturdy walleye, and the ever-elusive musky, all of which I had studied on YouTube and only one of which I had ever actually caught. Nor did we catch any of those exotic species on the trip in question, although we did haul in a torpid rock bass or two. But largemouth bass we caught in glorious abundance. We caught them everywhere we fished: beside docks, on sandy flats, from submerged grass beds, you name it. But most especially we caught them in a beautiful, winding channel linking Crawling Stone with Fence Lake. We would head out before dawn, hitting the mouth of the channel at first light, working our way back and forth between the two lakes. The scenery was breathtaking and the fishing was phenomenal. We were using Wacky Rigs with three- and four-inch worms, catching hard-fighting one- to four-pounders one after the other. In the evenings, we would sit on the dock outside the cabin, drinking cold beer while pulling in a few more bass for good measure. Glenn was so happy that one evening he sampled our host's bottle of bacon-flavored vodka. The look on his face as his tastebuds rebelled was priceless. That whole experience was a treasure that I will remember as long as I live.

As I write this, it is early spring. The trees are beginning to bud, squirrels are raiding the bird feeders, and the water temperatures are rising. As is fishing fever. Grab your secret weapons, my brother: time to double up again.

A TWELVE-HOUR TOUR

Some years ago, I was fortunate enough to publish my memoirs,[1] a series of stories chronicling a rather unlikely life journey from government cheese dependency to a career in academic archaeology. One of those stories was about two of my favorite subjects: fishing, and my brother Glenn. The following text is excerpted from that book, with sincere thanks to the publisher.

One positive outcome of my changed life status[2] is that I reconnected with my siblings and their families in ways that were long overdue. One weekend I reconnected in ways that might very well have led to an early demise. Glenn asked me to accompany him on a trip to Milton, Florida, where our cousin Ted had an old deck in his yard needing dismantling. If Glenn and I did the dismantling, Glenn would get the lumber for free, which he planned to use for a deck at his house. Having lurked in near-seclusion for months due to the coronavirus outbreak, I was more than ready for an adventure. Knowing Glenn and Ted, I strongly suspected that misadventure was to be a more likely outcome. I was righter than I knew.

We left Glenn's house early one morning in his battered old Ford pickup, a vehicle distinguished by a lack of air conditioning and a set of bull's horns prominently affixed to the hood. An equally used twelve-foot trailer bounced along behind us. Somewhat to my surprise, we got to Milton without mishap and spent the next day working furiously, drenched with sweat in the early September humidity of the Panhandle. The lumber was quite weathered, so screw heads instantly stripped when we tried to back them out with a drill. Instead, we had to pull things apart via a combination of hammers, crowbars, brute strength, and a Sawzall when we could fit the blade between members. Once dismantling was complete, we had to use a jack and chain to pull the four-by-four posts out of the ground, following which the encasing concrete had to be busted from post bases with a sledgehammer. Finally, all the encrusting hardware (joist brackets, squirrel feeders, and so on) had to be wrenched off and the lumber loaded onto the trailer. We also loaded up an old motorcycle and some kind of enormous tractor part, so the combined weight of the payload was considerable. It was a long day, but great fun, and the cold beer went down very well that evening. We went to bed weary, satisfied, and in eager anticipation of the all-day fishing trip planned for the morrow.

We got up very early the next morning, intending to hit the water by first light. Ted, that generous soul, had bought lots of bait the previous afternoon: frozen fish, packaged squid, and live shrimp. We stopped at a gas station for final supplies and headed out, arriving at the launch on the Blackwater River right on schedule. Ted's wife Star, Glenn, and I all took our places on the boat, one of those really large, double-pontoon things with a nice, canopied deck. Ted expertly backed it into the water. Having little draft, the boat floated easily off the trailer, and Ted went to park the truck. The plan was for Star to ease the craft over to a dock, where Ted would come aboard. It was a good plan, contingent upon only one key factor: the motor starting. When that failed, a premonition began to grow that the uncustomary smoothness of the operation up to that point was bait that Fate had used to lure us into a bleak oblivion. Ted strolled down to the water, and asked what the problem was. Star, her voice growing louder and more strained as we receded, explained the situation, which all of us pondered as the boat picked up speed in the strong current, our captain and the dock growing smaller by the minute. There was nothing else for it. Off came my shirt and shoes, out came my wallet and cell phone, and into the water I went. I gripped a rope in one hand and swam as best I could against an insistent current with a very large boat in tow. I finally reached the dock, blowing like a lovestruck manatee. I caught my breath while Glenn and Ted diagnosed the situation: a dead battery. Glenn and I fished from the dock while Ted and Star went off to buy a replacement. About an hour later the new battery was installed, the motor cranked easily, and we belatedly headed downstream toward Pensacola Bay.

The day was glorious: warm but overcast, the biting flies the only distraction as we slowly cruised down the river and into the beautiful bay. I hooked something big on a trolling line; it tossed the hook before we could get a look at it, but we took it as a good omen. We made our way to first one bridge then another, stopping at each to drop lines. We caught a number of small bullhead catfish, but no specks, snook, snapper, or other desirable species were to be had, probably due to our complete lack of knowledge where saltwater fishing was concerned. It didn't matter, especially as one o'clock pm was officially the start of Happy Hour. We eventually made our way all the way to the mouth of the bay, where we docked to use the facilities at a restaurant and where I discovered that I had no idea about how to tie a large boat to a wooden post in choppy water. I hugged the post till the others returned, then went to take my turn. Rest stop completed, we headed back upstream.

The day had grown long, and Ted remembered that an open bridge we'd passed was due to be closed, blocking access for the evening, so he told us to hang on, and opened 'er up. That boat could move! We clung to siderails, laughing and enjoying life as only slightly buzzed grownups can. Until, somewhere around the interface of East and Blackwater Bay, the engine cut off. The gas gauge read full, which was odd given that we'd been out all day. It became even odder when Ted mentioned that he'd not put any gas in that morning because of the gauge reading. It became odder still when we learned that he'd had the boat out a couple weeks before, for eight hours. One might reasonably have ascertained that two full days of boating should have dropped the fuel level some appreciable level, but with the gauge reassuringly indicating otherwise, like the loaves and fishes, the assumption seems to have been that we were good to go. That such was not the case became pressingly apparent as we began to drift back toward the Gulf of Mexico.

What to do? Our first gambit was to grab phones and search the internet for a boat towing company, although business hours had come and gone. We finally found a number that answered. Ted asked if they could bring us out some gas. "Sure, sport," the miscreant on the other end of the phone answered. "For eight hundred dollars." Stunned, we looked at each other as the seriousness of the situation sank in. At about that time, we saw a small ski boat with a single occupant heading in the direction we needed to go. Several loud whistles and exaggerated gesticulations later, he came over and gaped in astonishment as we explained our predicament, while the shore slipped slowly by and the sun nipped at the trees. After some consideration, it was decided that he would give us a tow, as he'd put in at the same landing. Relieved, we tossed him our mooring rope. As we did so, Star perspicaciously noted the rope's thin circumference and aged state. Within twenty yards, it broke. Fishing around, she found a ski rope in some recess and we hooked up again. It worked! Our benefactor turned his craft north and off we went!

For a bit. We had miles to go, and twilight was upon us. The small ski boat was laboring mightily, but we were only making about two knots. Smoke began to pour from our benefactor's motor, and the back of his craft kept dipping dangerously low into the water, which lapped hungrily at the gunwales. The futility of the effort quickly became obvious, and when a tiny boat ramp was spotted on the western bank, we headed that direction. As we neared the ramp, the water became so shallow that even the ski boat was hitting bottom. A new plan was hatched: we disengaged the towing rope and tossed out our anchor. Glenn and I stayed on board while Ted and Star clambered down onto the ski boat. They took positions on a flat platform near the bow; the skipper passed them both a beer; hope dawned. Then he opened 'er up, and they took off upstream, Ted and Star bouncing on the platform like ping pong balls, beer and river water spraying into their faces. Ted, I should mention, was recovering from a major hernia operation, and the blood drained from my face as we saw them clinging on for dear life, jostling like they were on a used carnival ride. Hoping that his innards would make it back intact, Glenn and I cast out our fishing lines, trying to relax with a bit more recreation before the light faded completely.

Which happened more quickly than anticipated, given the enormous, terrifying storm that came hurtling out of the east like some gigantic harbinger of doom, a rolling, black cloud bank sporadically illuminated by lightning that flashed and crackled across a front stretching in either direction for as far as the eye could see.

"Might get some weather soon," Glenn remarked casually as he flung his line out again. The words hadn't left his mouth before an incredibly cold blast of air hit us directly in our faces, a startling sensation given the heavy, humid atmosphere that had been blanketing us all day. In a strained voice, I suggested that we might perhaps consider reeling in our lines and taking at least nominal cover beneath the canopy, which was looking more and more dubious as shelter. Grudgingly, Glenn complied, and not a moment too soon. The storm hit us like a freight train, sheets of heavy rain driven almost sideways by powerful winds. The surface of the estuary looked like something from the Roaring Forties, all choppy waves and foam. And although the canopy did offer some protection, the wind splattered us with rain in random gusts, so that we both were thoroughly soaked in no time. I held my hand over the pocket that held my cell phone and tried to make nonchalant remarks as the storm grew ever more intense. Glenn made some less than nonchalant remarks when his side of the canopy suddenly folded and sent a generous freshet of water right down the back of his shirt.

It was very fortunate for us that the wind was blowing west instead of south, as I had no interest in visiting Guantanamo Bay, and as the anchor proved hopelessly inadequate under the conditions. The pontoon boat was bucking up and down as we were driven toward shore, straight toward the riprap lining the little concrete boat ramp. Fortunately for the boat, the wind veered southwest at the last minute, so that we missed the rocks and instead found ourselves beached on a parcel of land festooned with "No Trespassing" signs. Having been on the boat all day, I waded through the surf and urinated against one such marked tree, figuring that the rain would quickly eliminate any evidence of my crime. Then we sat on the boat and waited. And waited. And waited. Ted and Star couldn't find us, and the storm was so bad that they couldn't get sufficient signal to do an internet search for the landing location. When they could, the landing, minor cultural feature that it was, did not show up. Finally, they got through to us, and I waded out again to discern and relay the closest street names. By the time they rolled up in the truck, it was far too late to attempt any sort of loading. Ted knocked on the door of the house standing on the property where I had peed, which turned out to hold a completely understanding and sympathetic woman who graciously allowed us to leave the boat there for the night. Exhausted, we roped it to a power pole and drove the considerable distance back home to peel off our soaked clothing and crawl into welcome beds.

The next morning, we were at the landing early, and fortunately the tide was in, so that the boat was floating. Ted and Glenn waded out to the boat and poured a couple of gallons of gas into the tank. It cranked right up, and Ted was able to back the trailer down the ramp far enough to allow us to load. We were back home at a respectable hour, which was fortunate, since the coolers still held bait fish, long-dead shrimp, and tasty squid bits floating around with packages of pallid cheese, half-drunk bottles of water, and a remnant beer or two that looked oddly enticing despite the clinging bio-slime and the early hour. The stench was appalling. All the fishing rods were hopelessly tangled together, and the paper garbage bags had melted, but at least the gas gauge still read full. With inordinate cheerfulness, we cleaned the whole mess up and then spent the evening devouring some delectable grilled fish and vegetables Star whipped up, drinking nominally de-slimed beer and shooting the breeze in a tent in Ted's back yard as more heavy rain settled in. Life was good!

The next day, Glenn and I loaded up our belongings early and started on the long journey back to Mississippi. About a half an hour down the road, I realized I'd left my car keys behind, so back we went. That detour wasn't helpful, given that the rain was growing heavier still, the leading edge of Hurricane Sally churning her way up the Gulf. The consequent low visibility added considerable danger to the enterprise, as of course the trailer lights did not work. But an hour or so north of Milton, the skies cleared, everything on the trailer was holding well, and the tension slowly drained from my body. We could go only about fifty miles an hour with that load, but as long as nothing went wrong, we would make it back to Glenn's place well in advance of the hurricane. We both grinned as we passed the Mississippi state line and turned north. Homeward bound!

Until about thirty miles south of Meridian, when, traveling up Interstate Highway 45, a rear tire blew out on my side of the truck.

Never a good thing when you're hauling a large, heavily loaded trailer, particularly on a highway notorious for eighteen-wheelers. Ever unflappable, Glenn managed to slow us down and get the truck and trailer just off the highway, onto a tiny patch of asphalt where we could set a jack. We first tried a twenty-ton hydraulic jack that Ted had loaned to Glenn. The truck began to lift . . . and then settled back with a sigh, as the ram descended into the jack body while oil, or hydraulic fluid, or something else I didn't want to touch squirted out of what I assumed was the jack's mating orifice. We then set a frighteningly gracile scissors jack beneath the rear axle. I grabbed the ridiculously long crank handle, not wanting Glenn to undertake the operation as he already had undergone a number of previous operations; eleven hip surgeries, to be precise. I cranked and cranked, and slowly, *slowly*, the truck began to rise.

I was astonished at how weak I seemed to be, until I realized that I wasn't just raising the truck; I was raising the truck and the attached, heavily loaded trailer. The spare tire was beneath the bed, trapped beneath some ridiculous mechanism that had to be manually unscrewed, which meant that Glenn had to get underneath the truck. Wisely, he placed a second scissors jack at the back of the vehicle, as the primary instrument had begun to bow outward toward one side. By that time, we both were soaked with sweat, the only physical relief coming from the snatching winds that sucked at our clothing each time an eighteen-wheeler thundered past. I held my breath until Glenn finally got the tire free. He then tied the holding mechanism to some point beneath the bed with a piece of rope. Once he was clear, I emplaced the spare and threaded lug nuts back onto the waiting bolts. I eased off on the jack, the truck came down . . . and the spare immediately went almost completely flat. I cursed. Glenn chuckled, then rummaged around in the tool box until he found an air compressor that looked like it might have done duty at Verdun. He plugged it into the cigarette lighter and flipped the switch. It didn't work. He slapped it. It didn't work. I began to feel panicky. Glenn remained cheerfully unperturbed. He slapped it again, and grudgingly, with an aggrieved whine, it came to life. Waiting for the tire to inflate was like watching corn grow, but after about thirty minutes it held enough air that we considered it safe to head to the nearest town and buy another spare for the journey home. The station attendant whipped a used tire onto the rim in record time. Glenn gave him a pocket knife as a tip.

At that point, I was determined that we should travel via back roads rather than the interstate, which was dangerous at the best of times and downright terrifying given the load we were pulling and the questionable reliability of our conveyance. I would rely on Google Maps™ to navigate. That worked fine for a while, until my phone ran out of juice and I discovered that the charging cable coming out of dash wouldn't fit the receptacle. No problem. Glenn barked commands into his phone, which willingly began to issue directions. Which would have worked well, except for two complicating factors: 1) it was hot, and without air conditioning we had to ride with the windows down; and 2) Glenn is deaf as a post without his hearing aid, and possibly deafer than a post when, with hearing aid in, the rushing air coming into the truck blankets his aural receptors with an unbreachable wall of white noise. "Turn right on the next road," the phone helpfully suggested. Past the next road we went. "Make a U-turn and then take your next left," our electronic guide remonstrated in an accusing tone. On we sailed. I had the pleasure of viewing parts of the state I'd never seen before ere we finally, finally pulled into Glenn's yard, tired, hungry, but whole, with never a lost board or tractor part. Despite—or because of—the misadventures, the whole trip was enormous fun. What stories lie around the next bend in the road, I can't imagine.

REFLECTIONS ON THE WATER

My biggest worry is that when I'm dead and gone,
my wife will sell my fishing gear for what I said I paid for it.
—KOOS BRANDT

When I was eighteen years old, Granddaddy took me fishing on Grenada Lake. He was not well. He did not speak of it, but it was painfully evident when we stopped at a seedy filling station to get gas for the outboard. He pumped the fuel into a big, red plastic can, waving off my help but moving stiff and slow as he hefted the container into the boat, as if pushing against thick cotton padding stuffed into his coveralls. But there was no padding, as evidenced by the purple bruises petaling the skin of his thin arms like flowers in a Japanese watercolor. What he was pushing against was not something that was going to yield. I tried hard that day to make a show of fun to reward that sweet, mischievous old man, that happy, boozy superhero in the orange, possibly purloined, coveralls. I tried hard, but I doubt that I succeeded. I am not sure that Granddaddy would have noticed in any event. His powers had faded; like the comic book heroes of the Golden Age, he belonged to an earlier, more colorful era. We nosed around cypress trees, dropping minnows beside the trunks, but the water was quiet that day. At least Granddaddy did not have to clean any fish. A more bittersweet ending would have been him offering me his tackle box, which I would mournfully but bravely have accepted. Instead, he may not have even known which one of Bill and Joye's brood I was. I was just one of the "heathens," as he like to refer to his many grandchildren. But it did not matter; we heathens respected age, even if we did not always revere it. Now, in my seventh decade, I find it right to share this memory of Granddaddy. My Granddaddy, who loved to fish, and who pushed back against the inexorable current of time to take me with him that day. I thought he would live forever.

None of us does, of course, which is just as well. If you lived for too long, memories would become more like mile markers and less like old friends. Best to check out with something like grace, taking those old friends along with you. We all bear joy, we all bear sorrow, mostly thanks to one another. If we all could go back, what would we be going back to? Something better? Probably not. Whatever else we may be, we are creatures of the moment. Somehow, we forget that all the time.

Such thoughts were constantly on my mind in the years after Janet's passing. Thoughts about life, legacy, mortality. Questions without answers. Insomnia was a frequent companion. Fortunately, it pays to be up early when fishing, although not all of my fishing partners were equally enthusiastic about that idea. I dragged them out anyway, and the stock of old friends grew along with the tally of glorious sunrises, always a welcome reward. I did not yet know my way, but when fishing, the journey and the destination were the same. Fishing helped sustain me until, finally, a morning came when I awoke and was myself again. A different self, of course, but I knew that I could move forward at last. And when I did, the path led to something unexpected. It led to love.

There are marvels in this world. Josie is one. Smart, funny, beautiful, kind, generous, insightful, decorous; dauntless, resilient, assiduous, impatient, roguish, fiery, and crass. She is all those things, and more. I said at the beginning of this book that fishing somehow puts me in touch with who I really am. And so it does. But not like Josie. She attaches me to so many things. This moment of ours is special. Our house lies in the lightly settled Tombigbee Hills east of Tupelo. The Tater Hills, according to the old-timers. Only fifteen minutes into town, and we can hear the train whistles blow at night. How sweet is that? It is just a five-minute drive to a public lake where I have watched Josie cast from the dock. That is a moment that should live forever.

The Earth carries on in its elliptical journey around the sun, a passage marked in my mind largely by seasonal variations in bass behavior. Marked also, of course, by the inevitable changes one experiences with age. The time is fast approaching when my nickname may again be Mount Baldy. But I have not given up on Captain Awesome. The fishing adventures continue. Josie is warmly supportive of my outings, although she was compelled to point out that if I can remember people's PBs, I ought to be able to remember birthdays. Keith and I still cull the Barrel upon occasion. He finally made the move up from Barbie pole to Zebco. Keeps me on my toes. Bobby enters a Zen state with his purple jelly worms whenever we hit the water. My friend Jimmy, a.k.a. Peckerwood, routinely outfishes me using a plastic salamander, the rogue, but he takes me to the Church Pond, where heavenly lunkers lie in wait, so it is all good. Friend Billy is reliably waxed, but that never dims his crescent-moon smile. What a joy it is to fish with all those folks.

Andrew, a good friend and fellow archaeologist, invited me to come fish for striped bass in Tennessee. Stripers are anadromous, spawning in freshwater. In the mid-twentieth century, states began stocking them in public reservoirs for sports fishing. They are incredibly strong fish that can exceed five feet in length and seventy pounds in weight. I had only ever caught one striper in my life, during a trip to the Grenada spillway with a bunch of fellow roofers back in the early eighties. One of the guys was even more incongruous in that crowd than I, a long-haired, gentle, hippie sort who seldom drank or contributed to the quasi-religious exchanges over this or that college team. A real standout, in other words. The Fishing Gods saw fit that day to reward his genteel behavior, as he brought in striper after striper, some exceeding ten pounds. I, meanwhile, was failing to get even a bite. Finally, I felt a strike and set the hook, excitedly yelling at my companions to get the net, get the net! Their reaction when I pulled in a half-pounder can be imagined. Of course, they insisted on using the net. But at least I did not get skunked.

I expected better luck with Andrew, who is a true master, making his own lures and traveling around the country to seek out the monsters. We left his house well before dawn, sleepily reminiscing about archaeological victories, losses, and general gossip—ours is a small community—while waiting for the bitter convenience store coffee to eat its way into our systems. We hit the spillway at dawn. There was only one other person casting, upstream and much closer to the dam than we were. I would have been envious except that he caught nothing while we were there. We commenced casting out Andrew's beautiful homemade plugs with what must have been surf reel combos, enormous rigs with heavy braided line, retrieving the lures with repeated snatches against the current. It was real, physical work, and great fun, except that nothing was biting. Andrew was reassuring. I changed to a shallow-running plug that generated a pronounced rippling feeling as I pulled it back against the current. That meant action: action produced results. The strike was unlike anything I had ever felt, a startling, savage jolt like the lure had been hit by some formidable denizen of a Mesozoic sea. I finally managed to land that fish, a fourteen-pound striper, still my PB for the species. Maybe for any species. Andrew observed that they got a lot bigger. I obliged him by hooking another one a few casts later, a *BIG* one, that literally pulled me down the rocks toward the water. Its strength was astonishing. Probably a twenty-pounder. My adrilogen-flooded brain abruptly decided that I should mimic the scene on TV where they hook a marlin out on the briny blue: haul back, drop the rod tip while reeling up slack, haul back, repeat. The fish flung the hook the instant I dropped the line. Darn brain. I was not crushed, though; I was proud for that fish. And for so much more. Andrew, too, has known loss. Andrew, too, has found solace on the water. Thank you for sharing, my friend.

Josie's Uncle Jimmy lives out in the country between Oxford and Water Valley. With his rugged features and big frame, he looks like he was drawn by Jack Kirby. I call him Skipper; he calls me Little Buddy. Uncle Jimmy won an everlasting place in my heart by gifting us with a vintage, 1970 V-hull Duracraft boat. She has seen some wear and tear over the years, but not much; she is one tough vessel. The faded, camouflage color scheme did not do her any favors, but a new paint job and a custom seat installation worked wonders. She is beautiful, elegant, and stable, moving effortlessly through the water under the power of a trolling motor, cutting a respectable wake under outboard. Skipper suggested that I call her the *Minnow*. I opted for the *Josephine*. There are marvels in this world.

Uncle Jimmy also is a fisherman, of the resolute sort that could navigate the Arctic wastes with nothing more than a sharpened stick and perhaps a satchel of ketchup packets or a bottle of McClard's barbecue sauce to add a little flavor to his char. Mississippi waterbodies being more clement than Arctic wastes, he likes to wade out among the trees lining the more inaccessible parts of the large reservoirs in our part of the state, jigging for crappie while nonplussed moccasins cock curious eyes at the intrepid interloper. He once took me to see a favored fishing hole not far from his house. I am no stranger to the outdoors, but that particular expedition lay somewhat outside my prior range of experience. We climbed onto his Gator and crossed the blacktop road in front of his place, heading down a dirt track into the broad bottomlands of the Yocona River. We powered through muddy fields and across yawning erosional gullies, Skipper pointing out various landmarks along the way while I sank fingerprints into the metal roll cage. We experienced a lot of what in the airline industry is called "turbulence." One bounces on those seats. One substantial creek had to be crossed, the steep trackway being lined with sizable chunks of riprap on either side of and fully across the drain. A good way to stem erosion while allowing for healthy water flow and effective biotic transport, I thought in end-of-life fashion, as we lurched and bounced along like a lunar buggy. Skipper paused while I got out and removed one particularly large boulder from our path, following which we lurched and bounced our way up out of the channel, two human concertinas expanding and contracting as recoil and gravity played a duet with our spines. It was a bumpy ride.

The closer we got to the river, the wilder the landscape became. Eventually we reached the spot from where, Skipper informed me in all seriousness, he launches his boat. Even in my state of recovering equilibrium, I had to wonder, how? There was no ramp. No vehicle trackway was discernible. Without being elevated by the Gator, I am not sure I could have seen much of the water through the thick bankside foliage, green leafy stuff loosely troweled into a skeleton-frame horsetail superstructure. I stood on the seat for more than just a vantage point, as such wet-ground flora provides prime real estate for our more sinuous neighbors on this Earth. The view was worth it. The scene was astonishingly beautiful, a primeval looking oxbow fringed by thick vegetation, dead trees rising out of the water, quiet and still except for the humming, sub-frequency pulse of life. It would not have surprised me one bit to see a brontosaurus lift its muck-draped head up out of that swampy hole. I cannot wait to fish there with Skipper. It will be legend.

One healthy change of recent years is that my tackle mania has declined considerably. I still buy one or two H&Hs a year, mostly out of nostalgia as they are employed mainly in the spring along with little minnow-shaped swimbaits and the ever-trusty Rapala. Summer bass fishing has become almost exclusively a thing of Ned Rigs and Wacky Rigs, with Whopper Ploppers and buzz baits added for the fall topwater bite. That is almost my complete repertoire. But it is always fun to pull out secret weapons when conditions will not yield to the ordinary. I may borrow Glenn's ridiculous stag beetle when he's not looking. I have more fishing rigs than I need, but not so many that things are out of control. I splurged recently and bought a Pflueger (Puh-*flueger!*) Patriarch reel paired with a Lew's Lite rod. It is an exquisite combo, beautifully balanced and with a lively action. The days of Henderson's red-and-white bobbers seem very far away.

I continue my armchair exploration of fishing culture, which never fails to entertain and sometimes to surprise. For example, what if your PB was not your biggest fish, but your smallest? Such is the world of Tanago angling, a term that sometimes is mistakenly expanded to refer to "microfishing" in general. Tanago fishing originated centuries ago in Japan, where a tiny species of bitterling by that name inhabits small streams and the innumerable agricultural channels enmeshing the land. It is a remarkable sport. As one topical website puts it, "In Tanago fishing, everything is shrinked or is made smaller in anything. The most of the catch is around 4 cm, and tackle is optimized for the size. But once an 8 cm fish is on the hook, it is like a giant in a small world. Then, there comes the thrill and the game of saving the line from breakage suddenly. Additionally, it is considered to be a great achievement to have over 100 pcs of fish on your palms, to praise the smallness of your fish and the number of fish you have caught."[1] Amen, brother. Bait is usually some form of glutinous paste, although small insect larvae also are used. The gear is fantastic; beautifully crafted, multi-sectioned little bamboo rods, tiny bobbers and hooks, line like gossamer. Setting the hook is a matter of the merest flick of the fist, but the timing and strength employed must be perfect. The fish are kept alive in rectangular ceramic or clear acrylic boxes until eaten grilled, bones and all, in a piquant marinade, or they are kept as pets. Microfishing is becoming increasingly popular worldwide due to the angling challenge it presents, a primary goal being to add to the list of species caught. It is a peculiar, fascinating venture. Glenn always tells me I am catching "embryos" anyway, so I may have a competitive advantage should we decide to give it a try.

I still fish with Glenn every chance I get. He knows my ways. I get a little overexcited when the bite is on and I break my line, get snagged, or otherwise foul things up, especially if I am behind on the leaderboard. My patience evaporates. I spit vile curses. I am sure I snarl. Glenn just says I "get a look" on my face. I bet. But then I watch Glenn, who in addition to his many hip surgeries has had multiple operations on his hands. How many? I asked. Five, he thought, but perhaps he was just counting on one hand, haha. I once saw him madly stirring circles in the air, his reeling hand a good inch from the handle it was supposed to be grasping. Those are the hands he fishes with. He is simply unflappable, taking life's vicissitudes in stride while being judicially sarcastic about the whole illogical endeavor. It is simultaneously ridiculous and inspiring, and I frequently find myself trying to emulate him. Glenn helped me knot my life back together when I became a widower. Sometimes I stop fishing to tie his lures on for him, even when they're biting. Because that is what brothers do. Our partnership has its hazards. I blanch as we forge toward the shore while he is bent over steadfastly failing to pick up a dropped hook with numb fingers while his downward posture translates to "High" on the throttle hand. He flinches every time I whip a crankbait past his ear on my back cast. Oh, ye of little faith. We each routinely spill blood by impaling ourselves on treble hooks. Whopper Ploppers are particularly unforgiving in that way. We fish anyway, of course, what with being Hill boys and all. Neither of us objects when the boat must be beached for private purposes. Glenn even puts a cup in my end of the boat in case the bite is on and the shore is far away. That is what you call a fishing partner.

April 26, 2025, was a sad day in nautical history. *Old Leaky* having exceeded any realistic use life many times over, Glenn decided to scuttle her in his pond, employing as a towing vehicle the *Pond Whore*, another badly used, ten-foot jon boat that he picked up at an equipment auction and named in my honor. Our inaugural voyage in that less than stately craft had come to an immediate halt when water came rushing in through a surprising number of slits in the hull. Following an emergency welding session, she hardly leaks at all. She is a proud namesake. Our plan for consigning *Old Leaky* to her watery grave was simple: tow her out to a deep part of the pond, pull the plug, and watch as she went down. As plans go, it was a good one. Not surprisingly, what actually transpired was another kettle of fish altogether.

Things began well enough, Glenn paddling us out while I held onto a rope tied to *Old Leaky*'s stern. We got her into position over just the desired spot, a deep part of the pond north of the little island where a pair of optimistic geese nest every year. Glenn gently sculled to keep us in place while I leaned over and pulled the drain plug. In came the water, pushing spinning patterns out of the thick mat of leaves, twigs, and other detritus slowly building topsoil in the bottom of the boat. From her nest on the island, the female goose watched with a critical eye. My brother and I exchanged knowing grins. Would not be long now.

Some ten minutes later, grins had constricted into perplexed frowns and the goose had begun napping. The boat was taking on water, alright, but so slowly! We had an afternoon expedition planned at a pond we had never fished before and were antsy to be on our way. More minutes passed; perversely, *Old Leaky* still rode high. Glenn decided to speed up the proceedings. We left our venerable old companion floating serenely above what was supposed to be her final resting place while I paddled us back to shore. Glenn clambered out of the *Pond Whore* and walked to the truck, returning shortly thereafter with a pistol, an angular, black metal thing of sinister appearance. Following his instructions, I paddled to a position parallel with and about ten yards away from *Old Leaky*. Glenn leveled the pistol and fired. The shockingly loud report echoed flatly off the water; the goose woke up; daylight instantly shone through an impressively large hole in *Old Leaky*'s starboard side. A hole a good two inches above the waterline. As I was about to point out that germane bit of information, Glenn leveled the pistol again. I had just enough time to get my fingers into my ears before he emptied the remaining bullets in the magazine. *Bambambambam!* Nine-millimeter shell casings flew out of the ejector port to bounce off my chest, gathering in a pretty little brass arrangement around my feet. The goose was nowhere to be seen. After twice obtaining verification that the gun was indeed empty, I paddled us over. *Old Leaky* drifted tranquilly, two holes now visible in her side. Both above the water line. Perhaps, I suggested, given the mission at hand, it would have been better to shoot through the *bottom* of the boat?

Glenn grunted absently in reply, studying the situation. He reached down and lifted up on *Old Leaky*'s portside gunwale. Water began pouring in through the two bullet holes opposite, which, in combination with the weak infusion still issuing through the drain hole, finally produced the desired effect. She rapidly began to fill. Hastily, I retrieved from my fishing vest a harmonica brought specifically for the occasion. As *Old Leaky* sank lower, I played "Taps" in dolorous tones, right hand wobbling in slow time to produce a truly mournful vibrato. The moment became suddenly poignant. Perhaps soothed by the music, the goose had returned. Glenn pushed down on the gunwale to administer the *coup de grace*. Water cascaded over the side, filling the boat. *Old Leaky* finally submerged, sinking to the bottom to provide a useful bit of fish cover. The memories remain. Would that all our endings could be so productive.

I need to wet a line with the other brothers more often. Dennis would be crafty, so that I soon would have to resort to skullduggery to wax him. Perhaps I would pass him an H&H in the package without bothering to point out that the hooks do not come attached. Bennie would reliably be waxed, but he would take it all in good philosophical spirit. Were Hardy still with us, he would present a formidable challenge, but manageable if explosives were forbidden. Robert would be tough, but his restless attention soon enough would wander to other endeavors. Andy would insist on throwing his beloved Tiny Torpedo regardless of conditions, which should insure my victory. But you never know. In fishing, as in the rest of life, things do not always go by the book.

One day in the summer of 2022 I went fishing at Glenn's Pond. I hung my H&H up in the limb of a pine tree that stretched out over the water, ultimately breaking the line and leaving the lure dangling like a Christmas tree ornament, out of reach. About six weeks later I returned to try my luck again. It was a hot, windy day. I was fishing the far corner of the pond when I noticed something splashing underneath the aforementioned tree limb. It was a bass, hooked on my orphaned H&H! The fish must have seen the lure moving with the wind and come out of the water to strike it. The chance of that occurring was miniscule; the odds of it transpiring just when I happened to be there again are so remote as to be ridiculous. I relate this story for two reasons: a) to show that, in fishing, anything can happen; and b) some fishing tales are true, no matter how improbable they may sound.

I have been blessed with a fortunate life. Time to start closing some circles. I would like to close one with this book. Come with me. Let's rig up a cane pole and head back to Elkins Creek to see if the bream are biting.

RETURN TO ELKINS CREEK

Everyone should believe in something. I believe I'll go fishing.
—HENRY DAVID THOREAU

A mile. How can it be only a *mile* from the Bridge to Highway 413? As the crow flies, anyway; longer if one follows the meandering creek, of course, but still . . . a mile? That was a journey of mythic proportions on skinny little legs, back when hot summer days under blue, blue skies seemed to last forever. I know the topographic map I am inspecting on my phone does not lie, but I am amazed all the same, especially when I think about how many fish we caught out of that short stretch of creek when I was growing up. I last fished there some half a century ago. It is now April 2025, Choctaw County, Mississippi, about four miles north of French Camp. So much has changed. I am here to discover what remains.

The old house does not, its shockingly fast decay after Mom's death in 2002 putting a full stop to any lingering childhood I might have harbored. Seeing that house in its abandoned state was hard. I felt like a stranger the last time I walked around inside its empty little rooms. How on Earth did nine people—sometimes more—live there? It was not salvageable, short of expenditure of more cash than any of us could afford, and with no one living there to take constant care of the place the money would have been poorly spent anyhow. The sad old thing eventually was torn down. A bulldozer had to be called in to bury the heavy concrete wheelchair ramp that had allowed Mom to continue residing at home, for a while. Flowers still grow there in silent lament, mostly daffodils taking little yellow dragon bites out of the spring air, and Pop's hand-built shop still stands, although one would be foolish indeed to enter the shaky old thing these days, especially given the number of copperheads that must reside underneath the cattywampus floorboards. The massive old oak trees also are long gone, the last one in the back yard cut down by Pop, who had convinced himself that it was dying and would demolish the house should it fall. No doubt, but when he finally managed to fell it via repeated applications of a chainsaw far too small for the job, it was solid inside, with no trace of rot or disease. A metaphor is there somewhere, but searching for it is too depressing. I loved those trees. And the earthworm-sheltering wrack that accumulated beneath them. It was good while it lasted.

My pole is storebought, a red, heavily lacquered thing of two parts that socket together via a metal sleeve, much longer and far more beautiful than the hand-cut yellowish-green cane poles of yore. The small cork is of red and white Styrofoam, recalling in color if not in shape the spherical plastic bobbers of my youth. The line is weighted with a small split shot and holds a little golden hook. I had considered capturing grasshoppers from the yard in front of Glenn's house for bait but decided against it. As a kid, I always hated the way they jumped and clawed against the glass walls of the mason jar, banging chitinous heads against the metal lid in an instinctual bid for freedom. And then there's the tobacco juice, the same revolting goop that polluted Nick Adams's hook in Hemingway's *Big Two Hearted River* and for which scientists employ an even more unlovely descriptor: "defensive regurgitation." No, thank you. Today I am using redworms, but not ones that I dug up. I am committing sacrilege by using storebought worms. My younger self would have been scornful. My younger self had better knees.

I am alone. I thought about asking my brothers to come along, but most of them live too far away to make such a whimsical outing worthwhile. Robert and Glenn are the only two still living close by. I figured that I could cajole Robert into accompanying me by telling him what I would write: "In earlier days, my brother's argumentative feet might well have decided to pitch him—and the bait—into the water, but nowadays he moves with the grace of a gazelle." But even gazelle have their bad days, and I need that bait. Plus, Robert never slows down. Today, I want to take my time. I talked to Glenn about it, and he rightly pointed out that eleven hip surgeries are not conducive to navigating the bluffs that awaited. He did, however, volunteer to have an oxygen tank waiting for me and a chain hoist on his tractor should I need an emergency extraction. That is how he gets his deer back to the house. Good ol' Glenn.

Every step along the road to the creek is a reminder of how different things are from those days of long ago. The road is now paved, sort of, with a thin, ugly cover of what must be the cheapest asphalt in manufacture. When we first moved into the old house, the road was dirt. It was a big day when they laid down gravel. Now this. And thus do our paths evolve. Huge thistles stand at strangely even intervals along the eastern road verge, heads bowed as though honoring a funeral procession. Drooping branches of some white-blossomed thing—Groundseltree? Sea myrtle?—reach protectively over the roadside ditch opposite but fail to enliven the scene. Where once there was pasture across the road there now is a wearisome plantation of stunted, starved-looking pine trees crowded far too close together, a dark, forbidding stand waiting forlornly for a thinning that may never come, the bottom having dropped out of the timber market years ago. Forested it may be, but it is a desert nonetheless, where life is concerned. I want to set a match to it. I move on.

The Bridge is not, anymore, that lovely old wooden structure having been removed years ago and replaced with an ugly, oversized concrete box that admittedly helps with the flooding but which lacks anything like personality. No place for trolls here; no magic, no welcoming perch, just a drab, soulless functionality. The creek itself is much diminished at this point, scarcely more than a shallow trench, the result of massive sediment dumps from poor logging practices upstream and heavy equipment shaping of the bank for emplacement of the unsightly box. No water here worth fishing, and the beautiful overarching trees have long since disappeared. There is no smell of creek at all. On the downstream bank lies another unlovely artifact, a plastic shopping bag holding what from a safely hygienic distance looks to be a pair of beige corduroy pants, sodden from the recent heavy rains. We have a word in the Deep South for people who toss out trash, abandon puppies, or otherwise do not give a damn about anything outside their own limited sphere. They are "sorry." I am sorry, too; sorry that I do not know who tossed that gross thing out of their car. I would return it with real satisfaction.

Before I plunge into the woods west of the road, I check the charge on my phone. I doubt I will need the oxygen tank, but the bluffs may be steeper than I remember, and although it is not yet high summer, snakes are a concern, so best to be in touch with the outside world. Besides, Josie insists. That woman has good sense. As do I, belatedly, as I am wearing boots rather than roaming barefoot as I once did. Shirted, rather than bare-torsoed; long pants instead of ragged blue jean shorts. I even applied some tick spray. When I was a kid, finding a tick on your person was an occasional nuisance; now, a simple stroll around the edges of your yard might bring in two or three unwanted passengers. Go spend a day in the woods, and you are liable to come back with dozens of "microticks" so small that they remain unnoticed until a persistent itch gives them away. This dramatic increase in arachnid fauna has been well documented by scientists, who uniformly attribute the phenomenon to climate change and who note an associated upswing in alpha-gal syndrome. Most folks around French Camp do not believe that climate change is real. But they do so love their red meat. How that disconnect will play out is anyone's guess.

Besides my long, storebought pole, which is going to be a major pain as I navigate the forest, I have one film canister holding hooks, another holding split shot, and my little carton of worms. No stringer, as I am not planning on keeping any catch. What I hope to land today are memories. There also are two small bottles of water comfortably couched in my fishing vest. As children, we simply drank from springs issuing into the creek, cupping the cold, cold water in our hands or spraddling a spring Spiderman style and putting lips directly onto the liquid surface. It was delicious, but these days I would worry too much about parasites. Opting for the microplastics instead.

There is an opening in the wood line. I bend down and pass through, holding the pole horizontally, butt-end in front, to minimize snags. Immediately, I am struck with wonder. The woods are just as I remembered them, lovely and cool. *Big* trees—white oak, red oak, hickory, beech, the occasional black walnut or solitary pine. Little understory, so that the first bluff rises up beautiful and easy to see in the open woodland. One huge old white oak has fallen, blocking the creekside path we used to follow. I will have to pass over the bluff to get to the first fishing hole. Before I do, I check the beech tree at the east end of the bluff, where P. P. once earnestly inscribed his 4-ever love for K. A. The letters are still there, if you know where to look, but unreadable, as bark has filled them in around the edges. Most wounds heal, with time.

No longer a reddish-brown, sandstone-pebbled line, the old trail leading up and over the bluff nonetheless is still visible, a shallow depression mostly covered in leaves old and new. It was far clearer back in the day when bare feet kept it so. How many hours did we spend playing on that bluff? How many times did my little legs pump me up and down the steep trail? I got my first squirrel there. Mom cooked it and a few others up with dumplings in a thick, floury sauce. We ate a lot of squirrel and dumplings in those days. I liked fish better, but there was nothing wrong with squirrel meat; at least it helped one to face the gooey dumplings. The best woods-fare was rabbit, enormous canecutters that bolted at high speed when jumped by a dog out in the flat, wet bottomland. You got one shot. That usually was enough with a 12-gauge shotgun and No. 8 shells. Panfried, those rabbits were unbelievably tasty. Did not go far, though, with nine or more mouths to feed. Two was a good day's haul for canecutters. As I became more practiced, four or five squirrels was the take on a typical fall morning. Supply and demand.

The creek is high, as it should be following the previous day's downpour. But not nearly high enough; the water in general is much shallower than I remember. At no time during the day will my cork be set for more than a foot of depth. That is a big change, one that does not bode well for fish. The sandbars have . . . too much sand. That is a weird thing to write, but it is true. They are festooned with black, black leaves, long dead but still supple from anerobic burial, only now exposed due to reworking of the sediments by high water. Maybe some of them will become fossils someday. The water is different, too, a color I have not seen before, like very weak tea with too much milk. The current is strong, requiring some strategy. I know to look for still water after a rain, in the quiet part of a meander bend or behind a log where fish can save energy while waiting for prey to appear. Especially still water bearing a scruffy blanket of creek foam. I come to the first such still place and stop to bait up. No foam, but the odd-colored water looks clean, surprisingly not muddy with fine sediments after the storm, which is an encouraging sign. I find it shockingly hard to thread a worm onto the hook. The numbness in my left hand makes a lot of things difficult, but *baiting up*? I may have reached the point where the clan would abandon me were it not for the vast store of useful traditional knowledge I carry about in my aging noggin. At least, that was my plan before the Internet. Now, I will just have to rely on my natural charm. Been nice knowing you.

I drop the line into the water, lifting and moving from spot to spot, reminding myself to be patient. Soon enough, the cork bobs a little, bobs again, then begins to sink. I pull up, too early, but get it anyway, a little yellow mudcat perhaps five inches long. It is barely hooked. I take a quick picture with the phone and release the fish back into the creek. Unknowingly, I have just released the largest catch of the day. I plop the cork back down again. A large minnow takes a nip at it, perhaps with amorous intent. A few more tries yielding no further bites, I move on.

The Log is still recognizable as a fishing hole, although the original barrier has been replaced by three others that have fallen into the creek, Pick Up Sticks style. Foam is piled up on the far side of the meander bend, just beyond the reach of my pole. There was a time when I would have scooted across one of the fallen logs to reach that choice water. That time has long since passed. I fish where I can, attention diverted by trying to identify old landmarks, but nothing is biting anyway. Then I notice a large, green bullfrog casually watching me from his perch on one of the semi-submerged trunks. I take his picture. He does not seem to mind. The gently rippling water is immobilized in the still life on my phone screen, making it look like he is sitting on the edge of a sheet of crinkled cellophane. I move on, so as not to disturb him. I am encouraged. Bullfrogs are voracious predators. They eat a *lot*. There is life in this creek.

Just beyond the Log is the second bluff, higher than the first and much larger than I remember, its sheer face rising straight from the water, ruddy clay loam sitting unconformably atop bone-white clay-stone. I will have to go up and over again. On top, I pause for breath beneath a gigantic old white oak, a tree that I must have seen a thousand times growing up. A smaller oak nearby sports a tree stand, accessible via a narrow metal ladder leaning against the trunk. The view of the open woods from that vantage point must be spectacular. Some forty yards further south I can see the bright, sickly green of a vast clearcut now overgrown with a tangled mantle of weedy vegetation. The contrast with the beautiful, mature hardwoods covering the bluff could not be more striking.

I descend the bluff to find a stretch of creek with relatively slow-moving water. I drop my line in, raising and repeating, raising and repeating. My eyes wander to the stately poplars rising from the bottomland across the creek. The mosquitos discover my presence, distracting me further. When I look back at the water, my cork is nowhere to be seen. I pull out my second catch of the day, a creek chub some three inches long, light brown on top, silvery below, with a faint yellow stripe between. Chubs were a rare catch in times past. I catch a couple more, and have several drop from my hook. Clearly, chubs are rare no more. I toss the ones I catch back and move to the next stretch of fishable water.

En route I stop to examine another artifact, a board, lifted by high water to be caught in some bankside shrubs. It is old, very old, blackened like the leaves by long submersion. Probably fourteen inches wide, but only an inch thick and about four feet long from sawn to broken end. There are a number of crudely bored holes in it here and there, and a squarish notch hewn into one side. It almost certainly came from the Tabor sawmill, a modest concern that once stood upstream between the Bridge and Bennie's Island back in the early twentieth century, when oxen served as skidders and enormous trunks were hauled individually on narrow, wheeled carriages. The site is infamous for a tragic accident, when sixteen-year-old Gus Boyd had his leg severed and died there on August 8, 1918. It has taken over a century for the board to make its way this far down the creek. One wonders if it is a board that he handled. I move on.

A still place, line in the water again, and the cork fidgets this way and that, sure sign of a turd tapper worrying the bait. I wait, and wait, but the cork never goes under. Finally, I lift the line and yet another chub rises out of the water. It is ridiculously small. The merest embryo. A chublet. More the idea of a chub than an actual fish. It releases its sucker-like grip on the worm and falls back in. Perhaps I would make a successful Tanago angler after all. With practice.

I come to the highest bluff in the series, a truly impressive hill rising perhaps fifty feet above the creek. I will have to climb it as well, as once again there is no way around at the base. I check my phone; no bars. Hopefully I will not in fact need the oxygen tank. The clearcut is still visible outside the hardwood buffer fronting the creek. It must be huge. Perplexed by the lack of bream bites, I lean my pole against an oak tree and venture into that cutover hell to search for grasshoppers. Not a one to be found in that horrible landscape, and it is stifling out there, a foul heat rising from the choked, leafy mess. I come back into the cool shelter of the woods, kicking over leaves to see if some delectable, scuttling insects might be found. Instead, I immediately see squirming earthworms, taunting me for having spent good money for bait. My younger self would have been amused. So is my present self, sort of. Live and learn.

At the bottom of the bluff is a shallow stretch of creek, water happily chuckling over the pale yellow sand substrate. What a beautiful sound it is. I could stay in this spot all day, listening to creeksong, but there's fishing to be done, and the mosquitoes have followed me. A little further along, creamy foam is piled up behind a fallen poplar tree. I drop my line and watch the cork disappear within the thick froth. I move it around, like stirring a cappuccino, until I can just see the surface of the water. In days past, bait dropped through a foam layer would immediately be taken. I fish there for some time without a nibble to show for the effort. Where are the bream?

Eventually I come to a truly impressive meander bend, almost looped back upon itself, where the water is slow moving and somewhat deeper than I have yet seen. A two-story house is just visible past the thin buffer of hardwoods and across a field on the far side of the creek. I am getting close to the highway now, hearing traffic on the road. More comforting is the ardent discussion taking place between two woodpeckers who seem oblivious to my presence. I drop my line in. The cork moves immediately in a pattern that sends familiar signals to my brain. This is no chub. It plunges under and I lift the line. The worm is gone. I rebait and try again. This time I hook and land a small sunfish, perhaps three inches long. It is extraordinarily beautiful, with a red tail, yellowish-brown side spots, yellow belly, and a striking series of orange and blue lines arching from eyes to the edges of the gill flaps. A juvenile longear, perhaps hybridized with a pumpkinseed? It flashes brightly in the water as I release it. Motion catches my eye, and I watch a small, black water snake sweep foam aside as it swims to take shelter in a wodge of leaves piled up against a stick. Raccoon prints wander across the sand bar where I am standing, and arcuate hoof prints mark where a deer made its way to the edge of the water. I am not the first visitor here today. That is encouraging. I fish for a while longer, rewarded with only a few more chub.

Nearly to Highway 413 now, the sound of passing cars louder. Just west of the highway is the parcel of land where Pop was reared, youngest of an enormous brood. He must have spent many hours playing on the creek with his siblings, never dreaming that sons of his own would someday enjoy the same privilege, albeit at some distance upstream. I never really talked to Pop about such things. We were not close in that kind of way. I do recall one story he liked to tell, about a wooden crate of Coca-Cola that bounced out of a delivery truck to fall by the roadside near his family home. Most of the bottles survived intact, an indescribable treasure for a young boy in those lean, lean times. He kept the bottles submerged in the ice-cold creek, making the enchantment last by limiting himself to one bottle a day. I love that story. I wish I knew more. I wish that we had talked about the creek and the games that he played there. I wish that I could have written more about his childhood adventures in this book. Just one more conversation that should have been.

I have gone as far as I can, the bottomland suddenly choked with gnarly, wicked greenbrier. The older stems have turned black and barbed-wire tough. There is no busting through that barrier short of some serious pain, especially with my long, storebought pole. No sign of the old beaver dam, alas. The creek ahead is wider and deeper still, perhaps home to the bream that I have been seeking. I catch and release a final chub then turn to retrace my steps over the beautiful bluffs, back to Glenn's house, phone bars, and a hot cup of coffee. Not a lot to show for my trip, but at least I did not get skunked.

The upper reaches of Elkins Creek have changed. I had expected that. What I had not anticipated was the ways in which it has changed. There once were so many fish in that creek that they made a significant and reliable dietary contribution to a hungry family of nine. In relative terms, the waterway is now depauperate. The once-plentiful bream are essentially missing, apparently replaced as the dominant species by greedy little chub, a phenomenon common to degraded streams. The main culprit doubtless is sedimentation, which has all kinds of deleterious effects on stream fauna, as fish eggs are buried, substrates become unsuitable for bedding, and so on. I suspect that the primary impact comes from high temperature, given how shallow the water was even after days of heavy rain when I visited. In high summer, it must be very shallow indeed. That shallowness, coupled with absorption of heat by suspended particles, equates to water too warm for many species.

And yet, I feel a sense of optimism. Perhaps it is a result of maturity, but increasingly I take comfort in the long-term view of things. The clarity of the water in the face of so much rain means that fine sediments are not pouring into the creek. The hardwood buffers, narrow though they are in places, are helping to stem erosion. Especially if the surrounding landscape is allowed to reforest, the coarser sediments will, in time, be flushed downstream, increasing water depth and producing a substrate more welcoming for the game fish of my youth. Such a positive change must have happened in the past given how prevalent logging was in the early twentieth century. The largest hardwoods seen on my trek are in the neighborhood of eighty to one hundred years old, based on girth, suggesting that they began to grow about the time that the early logging industry collapsed. I have no doubt that the surrounding ridge systems were largely denuded in those days, yet the creek was healthy and rich with life by the time we moved to the old house in the mid-1960s.

Perhaps by the time the plank from the Tabor sawmill, lonely testament to the unfortunate Gus Boyd, makes it another mile downstream, upper Elkins Creek will again be home to schools of hearty bream and chunky, chub-fed catfish. Perhaps other young boys and girls will brave the ticks and mosquitos to fish those rejuvenated waters, trying this hole or that, feeling the magic when the cork goes under, marching proudly home with a weighty stringer to share with the family come suppertime. Or perhaps this is all just the dream of an avid angler and lover of the woods who was privileged beyond measure to spend time on that creek when it truly was a place of wonder. Perhaps. But some dreams are worth dreaming, after all.

APPENDIX

A list of product names referenced in this book that have been copyrighted, trademarked, or registered is provided here. Readers are encouraged to try these fine products and/or to use the names as building blocks for haikus:

> Oh, Whopper Plopper,
> Churn the mirrored water's plane;
> Grant me a *baloosh*.

Arbogast Jitterbug
Bass Tracker
Beetle Spin
Canadian Mist
Culprit Worm
Duracraft
Flex Seal
Garcia/Mitchell 218
H&H
Heddon Tiny Torpedo
Hula Popper
Lance Captain's Wafer Crackers
Lews Mach Crush
Mercury Outboard
Mister Twister
Pflueger, Pflueger Patriarch Reel
Plano Box
Pocket Fisherman
Rapala Minnow
Rooster Tail

Shakespeare Barbie Fishing Spin Rod and Reel
Shakespeare Ugly Stik Rod
Senko
TikTok
Whopper Plopper
YouTube
Yum Dinger
Zebco

NOTES

I. LEADER LINE

1. Irit Zohar et al., 2022 Evidence for the Cooking of Fish 780,000 Years Ago at Gesher Benot Ya'aqov, Israel. *Nature Ecology and Evolution* 6:2016–28.
2. Sue O'Conner et al., 2011. "Pelagic Fishing at 42,000 Years Before the Present and the Maritime Skills of Modern Humans." *Science* 334 (6059):1117–21.
3. "History of the Fishing Reel." *The Fishing Museum Online.* https://www.fishingmuseum.org.uk/reels/overview.html.
4. See the appendix for a list of copyrighted and trademarked names referenced in this book.

III. FIRST CAST

1. "The Encyclopedia of Oklahoma History and Culture: ZEBCO." Oklahoma Historical Society, https://www.okhistory.org/publications/enc/entry?entry=ZE001.
2. ZEBCO. https://www.zebco.com/en/about-us/history.
3. B. A. Wells and K. L. Wells, "Oilfield Service Provider Zero Hour Bomb Company Produced 'Cannot Backlash' Fishing Reels in 1949." American Oil & Gas Historical Society, July 21, 2024. https://aoghs.org/oil-almanac/zebco-reel-oilfield-history.
4. ZEBCO. https://www.zebco.com.
5. Kevin Virden, *History of the Creme Lure Company 1949–1989.* Self-published, 2020.
6. Dyanne Fry Cortez, "Creme of the Crop." *Texas Parks and Wildlife Magazine,* March 2015. https://www.tpwmagazine.com
7. Steve Knight, "65 and Counting: Bass Fishing's First Soft Plastic Worm Celebrates a Birthday." *Tyler Paper Outdoors.* https://www.tylerpaper.com.
8. 1971 Creme catalog inside front cover. https://www.bass-archives.com.

IV. GEARING UP

1. *Louisiana Sportsman*, February 2023. https://www.louisianasportsman.com/
fishing/hh-lure-co-humphreys celebrate-60th-anniversary-of-proven-brand/.

2. Steve Miclette Jr., "H&H Lures Celebrates Sixty Years." *Texas Saltwater Fishing Magazine*. https://www.texassaltwaterfishingmagazine.com/gear/tackle/h-h-lures
-celebrates-sixty-years.

3. Ned Kehde, "History of Finesse Fishing." The "Original" Puddle Jumper.
https://puddlejumperlures.com/2016/05/19/ned-history-of-finesse-fishing/.

4. Dan Galusha, "A Legend—Gone, But Not Forgotten." Fresh Water Fishing
Hall of Fame and Museum. https://www.freshwater-fishing.org/legendary-angler
-virgil-ward/.

5. Rapala, https://www.rapala.com/content/rapala-general-information/our
-history.html.

6. "The Making of a Legend: How the Original Floating Rapala Transformed
Sportfishing Around the World." Rapala. https://blog.rapala.com/news/the-making
-of-a-legend-how-the-original-floating-rapala-transformed-sportfishing-around
-the-world/.

7. "The Birth of the Rooster Tail." Yakima Bait. https://www.yakimabait
.com/about/.

8. Travis Smola, "Heddon Lures: A Brief History of One of America's Oldest
Tacklemakers." https://www.wideopenspaces.com/heddon-lures-a-brief-history
-of-one-of-americas-oldest-tacklemakers/.

9. Joe Sills Sr., "Those Deadly Topwater Lures from the Past Still Work Today."
Fishing Tackle Retailer. https://www.fishingtackleretailer.com/twitch-coldwater
-twitching-smallmouth.

10. Todd Larson, "Fishing for History: The History of Fishing and Fishing
Tackle." https://www.fishinghistory.blogspot.com/2012/04/deconstructing-old-ads
-arbogast-hula.html.

11. Andy Arif, "The Universal History of Frog Fishing. Sorry, with Haunting."
https://www.andyarif.ro.

12. Antique Fishing Lures. https://www.mrlurebox.com.

13. "The First Frog." Bassmaster, April 21, 2016. https://www.bassmaster.com.

14. Don Wheeler, *The Frog Lure Collector's Guide: Frog Fishing Lures of Yesterday and Today* (Cincinnati: Whitefish Press, 2013).

15. Hugh Crumpler, "Splashing Bite." Land Big Fish. https://www.landbigfish
.com/articles/default.cfm?ID=422.

16. Bert Deener, "1,113 Giant Bass . . . and Counting." GON, April 27, 2010.
https://www.gon.com/fishing/1113-giant-bass-and-counting.

17. Or so it is often claimed. See discussion by August G. Costa and Amy
Fox, 2016, "An Experimental Evaluation of Gar Scale Arrow Points." *Houston Archeological Society Journal* 136:23–31.

V. CASTING INTO THE WIND

1. "Mississippi Distilled: Prohibition, Piety, and Politics Wins SEMC Gold Award." Mississippi Department of Archives and History. https://www.mdah.ms .gov/taxonomy/term/113.
2. Jerry Thomas, *How to Mix Drinks, or the Bon Vivant's Companion*, 1862. Dick & Fitzgerald, New York.
3. Mitchell Fishing Reel History. https://www.mitchellreelmuseum.com.

VI. OVERBOARD

1. Bama Bass. https:/youtube.com/@BamaBass.
2. Creek Fishing Adventures. https://www.youtube.com/@ CreekFishingAdventures.
3. NDYakAngler. https://www.youtube.com/@NDYakAngler.
4. Intuitive Angling with Randy Blaukat. https:/youtube.com/@ randyblaukatintuitive.
5. Hannah Barron. https://www.youtube.com/@HannahBarronOutdoors.
6. Hoodfishing US. https://www.tiktok.com/@Hoodfishing_Entertainment.
7. Rain Outdoors. https://www.tiktok.com/@rainoutdoors.
8. Kay.Fishing. https://www.tiktok.com/@kay.fishing.
9. Richard Gene the Fishing Machine. https://www.youtube.com/c/ RichardGeneTheFishingMachine.
10. Tackle Warehouse. https://www.tacklewarehouse.com/River2Sea_Whopper _Plopper/descpage-R2SWP13.html.

VII. DOUBLING

1. See chapter VIII.

VIII. TWELVE-HOUR TOUR

1. *Kudzu on the Ivory Tower: From the Backwoods to an Academic Career in the Deep, Deep South* (Tuscaloosa, Alabama: Borgo Publishing, 2021), 235–42.
2. Widowerhood. See chapter VI.

IX. REFLECTIONS ON THE WATER

1. Ty Hasegawa. https://www.tyhasegawa.com.

ABOUT THE AUTHOR

Evan Peacock is a former archaeologist in the Department of Anthropology and Middle Eastern Cultures at Mississippi State University. He is author of the acclaimed *Kudzu on the Ivory Tower: From the Backwoods to an Academic Career in the Deep, Deep South* and *Mississippi Archaeology Q & A* and coeditor with Patricia K. Galloway of *Exploring Southeastern Archaeology*, the latter two published by University Press of Mississippi.

www.ingramcontent.com/pod-product-compliance
Lightning Source LLC
Chambersburg PA
CBHW020237030726
47497CB00009B/3134